FROM A TO <A>

FROM A TO <A>

Keywords of Markup

BRADLEY DILGER AND JEFF RICE, EDITORS

University of Minnesota Press

Minneapolis

London

Published by the University of Minnesota Press
111 Third Avenue South, Suite 290
Minneapolis, MN 55401-2520
http://www.upress.umn.edu

Library of Congress Cataloging-in-Publication Data

From A to <A> : keywords of markup / Bradley Dilger and Jeff Rice, editors.
p. cm.
Includes bibliographical references and index.
ISBN 978-0-8166-6608-9 (hc : alk. paper)
ISBN 978-0-8166-6609-6 (pb : alk. paper)
1. HTML (Document markup language)—Philosophy. 2. Componential analysis (Linguistics). 3. Webometrics. 4. Metadata harvesting. 5. Sociolinguistics. 6. World Wide Web—Research. I. Dilger, Bradley J. II. Rice, Jeff.
QA76.76.H94F76 2010
006.7'4—dc22   2010019697

Printed in the United States of America on acid-free paper

The University of Minnesota is an equal-opportunity educator and employer.

16 15 14 13 12 11 10   10 9 8 7 6 5 4 3 2 1

FOR OUR GIRLS

CONTENTS

ACKNOWLEDGMENTS

This collection began as a panel at the Conference on College Composition and Communication in New York City in 2003. Along with contributor Michelle Glaros, Jane Love also participated in that initial discussion of markup. We thank Michelle and Jane for being a part of the original inspiration for *From A to <A>*. ·

We want to acknowledge the support we've received from friends and colleagues as we assembled this collection. We thank our contributors for their hard work. We thank Stuart Moulthrop and an anonymous reviewer for insightful comments and suggestions about the structure and content of the manuscript. And we are especially grateful to Douglas Armato and the University of Minnesota Press for encouragement and support of this project.

Introduction

MAKING A VOCABULARY FOR <HTML>

BRADLEY DILGER AND JEFF RICE

WE LIVE IN THE AGE OF MARKUP. As it becomes impossible to imagine a world without a World Wide Web, information organization, delivery, and production have converged on the simple principle of marking up information for given audiences. Markup, also called tagging, can be attributed to text, images, metadata, design, and behavior, affecting how readers engage with and encounter information. Markup bolds and centers, generates links, positions images on pages, produces output, and so on. Markup organizes and displays information; it institutionalizes practices relevant to online communication. Markup is the first thing Web writers engage, whether or not they imagine writing as tagging, coding, or marking up. In the age of new media, there is no way to avoid markup. Markup is text. Markup is communication. Markup is writing.

Unlike the Web, print culture has relied on markup indirectly, leaving most of its work to those who follow writers. Editors, publishers, designers, and printers all call on markup to prepare writing for production after writers have finished their work. Print has long supported this separation between markup and those who write. In the age of new media, markup is the process of writing, whether on a word processor, which hides the markup completely, or on the Web, where markup is easily accessible. We can no longer distinguish our writing from the markup we use to situate that writing within its given spaces. As Marshall McLuhan poignantly observed over forty years ago, "Typographic man can express but is helpless to read the configurations of print technology" (258). New media writers, on the other hand, read and write with the configurations of the new technologies. Markup is one such configuration.

To read a Web site such as CNN.com, NYTimes.com, or Washington Post.com is to read the configuration of the newspaper. To read a Web site such as News.google.com, ALDaily.com, or Slashdot.com is to read the configuration of aggregation. To read a Web site such as MySpace, Facebook, or any number of Weblogs is to read the configuration of social networking. Although all of these sites may assemble information as "news" to some degree, digital readers of such sites understand the different configurations at play, the different processes they participate in while reading or contributing to such sites, and the diverse ways each uses markup. McLuhan argued that the invention of typography eventually led to institutionalized practices such as the assembly line. Likewise, we can recognize that new media's engagement of markup configures a new type of institutionalized practice, one that configures a reading and writing public. We have not yet begun to discuss this public or the ways markup technologies affect its thinking and practices, such as displaying information in grids, using aggregation, rating participation, or showing personal updates. This volume asks readers to situate the reading and writing public within the framework of markup.

From the beginning, the Web has depended on the principle of marking up: making links, displaying images, adding metadata, and so on by adding a small amount of code to a plain text document. Tim Berners-Lee's WorldWideWeb browser set the standard for using markup that continues today with all contemporary Web browsers. A tremendous amount of early Web discussion centered on the roles markup should play online. On mailing lists such as WWW-Talk, contributors proposed new markup tags, discussed specific tags already in use, and debated browser functionality and compatibility. These discussions were, in effect, establishing the norms and rules of an emergent rhetoric of markup. Reading through the online archives of these discussions, we can witness the debates that shape practices, institutionalize thinking, and leave open possibilities for further work. We can see an emerging belief system as we read through conversations regarding font displays or image functionality. This belief system remains in flux today, as Web users and writers invent and reinvent new media practices: establishing the taxonomies of folksonomies; creating work-arounds for the inconsistencies of Cascading Style Sheets; complementing and blending others' work via mashups; struggling with increasingly problematic concepts of intellectual property; or creating dynamic interaction with languages such as PHP and approaches such as Ajax. These practices build discourses, communities, and spaces where markup shapes ideas, rhetoric, and communicative practices.

This collection argues that though WWW-Talk has closed, the discussion of markup is not over. We are still having it today. Berners-Lee extended a markup approach not only to the technical backbone of the Web, but also to its development, by shaping Web technologies through collectively marked-up specifications. Via Web-based institutions such as the World Wide Web Consortium (W3C) or via the more ephemeral spaces of social networking, discussion remains the model for the future of Web markup. *From A to <A>* extends the discussion further by bringing those initial conversations into current disciplinary vocabularies.

Markup, therefore, is not a static enterprise. Walter Ong wrote of the emergence of new media as the openness of the text, and markup exemplifies this process. Its growth and development accommodate the Web's accelerated expansion from a series of static pages to the emergence of the Web as an application platform. Despite the predominant roles markup plays in online writing, and despite the social, cultural, rhetorical, and technological implications of these roles, markup is often taken for granted as merely the code behind the text—constructed only by "Save As" or the work done by a given Web design staff. Browsers' "View Source" functions allow access to the code in most cases, but neither "View Source" nor any Web editor can reveal the other areas of experience markup effects. Indeed, "View Source" treats markup as simple code writers use in order to design Web spaces. Alan Liu makes such an argument when he demonizes markup, reducing it to merely another shallow design principle whose overall importance can be described by the throwaway term *cool*. Liu claims that "when we survey the coding possibilities added to each successive HTML specification," all we find is "a collection of tags and attributes (and style sheets) whose primary intent was to import New Typographical layout principles wholesale" (214). Wholesale importation, Liu argues, is done without reflection, critical thinking, or critique. For Liu, markup is the exemplar of new media's blame for the wide-scale replacement of identity, culture, ethics, and morality with a superficial fascination with technology. Markup, Liu argues, generates complacency.

Although Liu reduces markup to a simplicity, for many educators, markup is too difficult—best hidden from view, a representation of the almighty ghost in the machine. This position argues that markup is not worth learning because it is the least important element of Web writing. How-to books and Web sites promote this codeless vision of online composing. On the other hand, in hacker subcultures, markup is not powerful enough to deal with computing demands, as indicated by the hacker dismissal of markup as "not programming." Even Ted Nelson, who coined the term *hypertext,* argues that markup has devastated textual continuity when he laments that HTML's "beguiling simplicity pushes dozens of problems into the laps of users and creates a maintenance nightmare, resulting in the Content Management industry and millions of broken links" (170). Such positions, wherever they may be placed on the technological spectrum, ignore how markup constructs ideological, political, pedagogical, and other features of digital composing. Yet each position also attributes to markup a cultural significance by taking ideological positions—markup equates superficiality, is too hard to understand, is broken, or can be ignored—as well as the construction of ideological senses of agency—markup makes those who work with new media superficial, creates confusion, or is a waste of time. Instead of demonizing or romanticizing markup, we wish to demythologize markup as the mere coding of text or as the superficiality of appearance. Web editors such as Adobe Dreamweaver or the blogging platform WordPress, for instance, allow rapid movement, even simultaneous viewing, of code-based and WYSIWYG interfaces, reminding us that writing cannot function without code, and vice versa. These codes are both those formed of markup and those formed by culture.

Ong identified "the literate mind" as the interiorization of "the technology of alphabetic writing" so that various cultural phenomena are informed by logics and rhetorics shaped by literate practices (52). In the age of new media, in the age of markup, the literate mind has extended to the markup mind. We make this claim to highlight the relationship between technologies and cultures. Indeed, as the contributors to this volume argue, markup is the culmination of cultural experience: narrative, assertion, secrecy, pedagogy, identity, rules, materialism, agency, the body, and other features. In other words, notions such as literate mind or markup mind signify apparatus shifts, extended moments when, as Gregory Ulmer observes, institutional practices and belief systems are affected by technological change (*Heuretics* 17). The framework for inventing concepts such as literate mind or markup mind, Ulmer notes, is called grammatology, and our current grammatological shift, the convergence of literate and electronic practices, is what he terms *electracy.* Ulmer writes, "A goal of electracy is easy to state in this grammatological context . . . to do for the community as a whole what literacy did for the individuals within the community" (*Electronic Monuments* xxvi). Markup belongs within the overall notion of electracy; the consequences of its application are as important to literacy practices as the formation of print. Yet it would be too difficult to pinpoint a single cultural role for markup, or even for any one of its many tags. To do so, one would need a vast index, collection, or set of terms. *From A to <A>* does not offer a comprehensive index; instead, we approach markup through the genre of the keyword. Raymond Williams explained keywords as "the record of an inquiry into a vocabulary: a shared body of words and meanings in our most general discussions" (15). Vocabulary, Williams argued, revealed

> a history and complexity of meanings; conscious changes, or consciously different uses; innovation, obsolescence, specialization, extension, overlap, transfer; or changes which are masked by a nominal continuity so that words which seem to have been there for centuries, with continuous general meanings, have come in fact to express radically different or radically variable, yet sometimes hardly noticed, meanings and implications of meaning. (17)

The notion of vocabulary, in essence, is one of markup.

The keywords approach has become an important genre in academic writing: keywords of labor (Nelson and Watt), of the Web itself (Swiss), a particular writer (Stivale), and in specific disciplines (Heilker and Vandenberg) explore the vocabularies of these subjects. A keywords book presents the language of a body while articulating that language with any number of cultural, social, or rhetorical moments. As Williams describes his methodology,

> What I had then to do was not only to collect examples, and look up or revise particular records of use, but to analyze, as far as I could some of the issues and problems that were there inside the vocabulary, whether in single words or in habitual groupings. (15)

This collection both collects and analyzes. Its chapters gather together a body of markup in order to examine the vocabulary of new media writing. We situate *From A to <A>* within the tradition of keywords texts as well as within the emerging area of inquiry called software studies. As Matthew G. Kirschenbaum defines it, software studies "is, or can be, the work of fashioning documentary methods for recognizing and recovering digital histories, and the cultivation of the critical discipline to parse those histories against the material matrix of the present." In the recent *Software Studies/A Lexicon*, Matthew Fuller extends Kirschenbaum's definition to claim software as an "object of study and area of practice for kinds of thinking and work" (2). Software studies, as Fuller argues, introduces criticism to a software-based, textual body in ways that other texts (culture, literature, film, space) have enjoyed. "Software," Fuller writes, "becomes a putatively mature part of societal formations" (3). In his earlier foray into software studies, Fuller borrows from Gilles Deleuze and proposes the "blip" as means toward studying software and social formations. Blips, Fuller writes, are "events in software, these processes and regimes that data is subject to and manufactured by" (*Behind the Blip* 30). Events and societal formations, therefore, should foreground the various forces that make software cultural and programmable. Fuller's more recent collection, however, focuses only on the actions and objects associated with software: lists, loop, programmability, function, and so on. The chapter "Source Code" is the only part of the collection that recognizes code as an event. In its brief discussion of nonexecutable code, however, "Source Code" does not address markup. Missing, then, for a larger section of software studies is the relationship between software and markup. *From A to <A>* suggests that for software studies, markup plays a role as important as any other element associated with the digital. Our selection of the specific keywords presented in this collection gives a sampling of some of the most important societal formations and events generated by markup. As important as these terms are, though, for understanding software and culture jointly, this collection argues for a keywords reading that also understands each term's overall relationship to a larger body. In other words, markup is a body built from the vocabulary of each of its terms' vocabularies. It is a vocabulary of vocabularies.

In "Dictionary of Pivotal Terms," his rhetorical version of keywords, Kenneth Burke proposes casuistic stretching as a way to navigate the vocabulary of terms and their usage, as well as their various relationships to other terms. "Our proposed methodology," Burke writes, "to 'coach' the transference of words from one category of associations to another, is casuistic" (*Attitudes* 230). Burke argues that the haphazard nature of language causes this coaching process to produce rhetorical conflicts, disparate meanings, contradictions, and other supposed lacks in any given, used vocabulary. These gaps are where casuistry plays a role, for "coaching" meaning, Burke claims, results in "a firmer kind of certainty, though it lack[s] the deceptive comforts of ideological rigidity" (231). Coaching meaning stretches terms into associations, cultural analysis, critical gestures, and other related acts.

Similarly, rather than claiming a certainty for each keyword sketched out in this collection's essays, we recognize that each contributor to this volume caustically stretches each term's meaning in order to avoid the certain. The coached positions of *From A to <A>* are not meant as totalizing gestures or comprehensive analyses of the attributes of markup. Instead, following Burke's coaching metaphor, they tease out associations, meanings, and ideas as each writer leads the term along a specific narrative path. Overall, then, we see our contributors as rhetorical coaches who collectively build a vocabulary for markup by demonstrating the relationships between markup and the other vocabularies that, to borrow again from Williams, limit and pressure its work. Constructing this vocabulary leads our contributors to draw on a wide variety of conceptual fields and areas of inquiry—rhetoric, media studies, grammatology, philosophy, critical theory, design, and pedagogy, among others—examining implications and tracing the relationships for these fields simultaneously.

The chapters, we suggest, follow a trajectory of the Web page and its vocabulary. Thomas Rickert's "Tarrying with the `<head>`: The Emergence of Control through Protocol" tackles a fundamental law of HTML, one of its many enforced separations: metadata belongs in the `<head>`, and data in the `<body>`. This separation, encoded in Web pages by Web standards, brings to mind Cartesian dualism and Enlightenment rationalism, neatly hierarchical in both structure and mode of enforcement. But it also demonstrates the rise of the power structures of protocol. Drawing on the work of Alexander Galloway, Rickert shows how the shape of `<head>` is "part of the ongoing extension of protocol as the contemporary diagram of control in a posthuman world." Unlike the disciplinary systems it supplements, protocol asserts power via sets of simple rules with tremendous cascading effects. Protocol acts by recommendation and seeming voluntarism. As Rickert writes, "Indeed, part of the success of protocol is that it seems all to the good. Sure, one can take another direction, but why would one want to?" Notably, the head/body separation that is part of HTML standards forms the basis for protocols to call upon each other, through the underlying mechanism of headers that make up HTTP and other protocols. Indeed, the very process by which Web standards and technologies are developed, standardized, and implemented has embedded the logic of protocol throughout. Rickert concludes with an analysis of plagiarism in terms of protocol rather than morality, a move not intended to dismiss ethical considerations, but rather to explain the seeming irrelevance of those and other ideological forces more typical of liberalism than posthuman thought. `<Head>`, Rickert claims, operates like human subjectivity; it is a fissured, yet networked, protocol.

Sarah J. Arroyo seeks to reorient discussions of the `<b>` element away from the opposition of semantic and visual markup that has come to dominate discussions of `<b>` and similar supposedly nonstructural elements such as `<i>` and `<font>`. The typical reading of boldness—that it focuses only on appearance—carries arguments against visual culture into discussions of the Web and prevents thinking about boldness in other ways. Arroyo shows that a variety of other discourses are relevant: Debra Hawhee's reminder that classical rhetoric often calls on mind and body simultaneously; the idea of writing markup as conceptual design; and Hawk, Reider, and Oviedo's notion of "small tech" as the embodiment of the virtual. For Arroyo, moving beyond questions of the surface and the insistence on mastery fails to acknowledge the complexities of real-world communication, where depth and surface, semantics and appearance, and form and content exist contiguously. For markup, then, the physical rhetorical, real, and virtual collide, as exemplified by social networking's reliance on exchanges between surface (the network) and depth (the knowledge and rhetorical moves of invested users).

Colleen A. Reilly describes the `alt` attribute of the `<img>` element, originally intended to support the limited bandwidth of the early Web, but today a foundation of Web accessibility. `alt` makes Web content available to the widest possible number of browsing technologies. Reilly connects `alt=""` to the `alt.*` hierarchy of newsgroups and the unfulfilled promises often invoked in their defense—democratization and universal access. In much the same way that `alt.*` newsgroups were ignored (or blocked) by some Usenet providers, `alt=""` has not succeeded in creating an equivalent alternative to the visual Web it purportedly represents. Perhaps more importantly, `alt` (especially its low-bandwidth heritage) suggests that addressing disability must involve providing assistance to the less fortunate. Thus `alt=""` fails to acknowledge the complexity of (dis)abilities and represents them as unfortunate problems that can be addressed by a simple coding solution: the addition of `alt` attributes.

In "English `<A>`," Jeff Rice argues that links, the foundation of the Web, are too often seen only as simple connectors, vehicles for associative logics. Instead, `<A>` is the foundation of social spaces, an enactment of protocol that creates a superinstitutional network. Rice points out that Web applications' use of techniques such as tagging and folksonomy, in their rich application of `<A>`, are fundamentally different than Web sites' use of links as only connectors, a "click here" approach that models the Web as little more than a newspaper. Calling on the history of writing instruction in the United States, Rice proposes English <A> as an alternative to English A, Harvard's required first-year composition course, a pedagogical heritage that envisioned the writer as an individual learning about other individuals as well as individual subject matter. Instead, Rice uses <A> to rethink education as a pedagogical network, an assemblage (after Latour) of relationships.

As we note above, markup extends far beyond a single language (HTML) or a single form (the tag or element). Brendan Riley is the first of our contributors to combine attention to HTML tags with another form of markup, the stylistic codes of Cascading Style Sheets (CSS). Like Rickert, Riley begins with a separation encoded in Web standards, the division of Web coding into HTML for content and form, and CSS for presentation. Riley shows the separation as a desire for clear differentiation between aesthetic elements in a wide range of stylistic contexts: fashion, popular culture, writing style. But ultimately, there is little possibility of separation. The style guides one can find for any of these forms mix prescription and proscription, providing a strong hint that a clean break is not possible. Markup, too, belies separation, because in practice, developers must choose names for <div>, <id>, <span>, and other elements in HTML to ensure connections to their style sheets. For Riley, a better approach is recognizing these profound interconnections and seeing style as an action: Web writers, too, style their pages, making decisions about form and content, approaching questions of style and substance simultaneously.

In "An Accidental Imperative," Brian Willems writes about , the code used to create nonbreaking spaces, special units of punctuation that create spaces between words unable to break at the end of a line of text. Thus the invisible   has a dual nature: it simultaneously joins and separates. And like the language it joins–separates,   can be misunderstood or used incorrectly, giving it a menacing character. Willems shows this simultaneous presence of unity and disjointedness, of potentially menacing misunderstanding and error, as a form of pattern recognition at work in several different modes of writing, from the contemporary Web to Bruce Sterling's proposal for SPIMES, a form of extensible bar code that "is a gathering of flickering bits of information, the bringing of outside world knowledge into the inside of identity." Similarly,   can be understood, as N. Katherine Hayles argues, as a "flickering signifier" allowing the possibility for error and the sudden exposure of the (supposedly) outside world of the machine. The machine, in turn, sustains a core part of writing's identity. Willems's brief reading of Gibson's novel

Pattern Recognition shows the consequences of this flickering, menacing imperative for literature.

Several contributors write about markup that has fallen out of favor or that the W3C has tagged as deprecated—that is, not recommended for use even if still supported by Web browsers. Bob Whipple's essay on `<blink>` and `<marquee>` tells the story of two tags that, for many, were reviled from the start. But Whipple reminds us that the garish, over-the-top, flashy aesthetic these tags brought to the Web is a core element of American culture, as shown by Venturi, Brown, and Izenour's seminal *Learning from Las Vegas*. Labeling these tags as evil was a reaction to the do-it-yourself approach Whipple demonstrates: writers combining techniques they found cool and combining them freely. For these DIY writers, blink and marquee offered powerful and pleasurable self-expression; for professional designers, they exemplified unacceptable ornamentation and dilution of information (Nielsen), or the rejection of aesthetics and taste (Siegel).

Jakob Nielsen's attack on `<frame>` was direct and oft repeated: "Frames suck." In "`<frame>`ing Representations of the Web," Michelle Glaros takes up `<frame>`, another tag all but absent from the contemporary Web. As Rice points out for `<A>`, Glaros shows that `<frame>` exposed just how powerful literate visions of the Web remain: we prefer one document, clearly delimited and enframed, shaped by the stable narrative structures and forms of literacy, over the divided and disconnected viewport made possible by `<frame>`. Frames may suck because of usability problems, but those difficulties arise from our discomfort with nonprint, nonlinear, nonnarrative epistemology. Unlike film, where the fundamental disruption of the cut was refigured via montage as an fundamental unit of meaning, electronic writers have resisted similar moves for `<frame>`, insisting it remain in its literate form: a single, stable, all-but-invisible rectangle surrounding and supporting the content. "In its infancy," Glaros writes, "electronic writing promised to break these frames." That move may still be possible, if we can become comfortable with approaches to writing that, like installation art, invite the reader/writer to become a part of the artwork, disrupting the traditional approach where frames provide an unassailable border.

The collection's discussion of fallen tags concludes with the horizontal rule. `<hr>`, like `<blink>` and `<marquee>`, has nearly completely disappeared after its initial popularity. Matthew K. Gold notes that `<hr>` followed the function of rules in printing, helping writers break up tightly packed text into discrete units. As in print, Web-based rules began as simple lines but became more and more ornamented, eventually exceeding the capability of `<hr>` and replacing it with complex, often animated graphics. When `<hr>` was extended to allow colors and sizes, Web-based rules were transformed, becoming what W. J. T. Mitchell would call an *imagetext*. Not surprisingly, more ornamental approaches to these rules were roundly criticized as visual junk, echoing the attacks on `<blink>` and `<marquee>` Whipple documents above. The tag `<hr>` may be gone from the Web today, but rules remain a core part of Web writing: implemented in CSS, rules and lines guide the reader toward the content and enforce the linearity of writing.

How do bodies appear online? Jennifer L. Bay's examination of `<body>` continues the engagement of *From A to <A>* with the role the Web plays as a medium. Bay argues that even though well-respected critics such as N. Katherine Hayles and Mark B. N. Hansen have confronted the materiality of the Web, they have not done so for what might be called everyday Web sites—Facebook, CNN, Amazon, MTV, Ning, Ebay, and other locales where people reconstruct their bodies online as profiles. Supposedly, all content is in the `<body>`; is it also true that the body provides most content? In many ways, yes. Increasingly, on e-commerce and social networking sites, there is less and less identity play, less anonymous or pseudonymous access, and more projection of corporeal identity. Public rejection of fake bloggers and fascination with "authentic" reality demonstrates the drive to represent the "real" body in the profiles, avatars, and personas of online spaces—a sharp contrast to early models of online identity that suggested the corporeal was irrelevant. Instead, Bay, drawing on Bernadette Wegenstein, shows how the body (or even small parts of the body, such as skin or the face) functions as a medium. That is, via `<body>`, bodies are becoming content, a core part of the emergent rhetoric of Web writing.

Helen J. Burgess writes of the additions of embedded scripting languages such as PHP to acts of reading and writing markup. Burgess connects markup in general, but particularly scripting languages, to the use of pecia (printers' marks). Pecia, a literate form of markup, helped printers assemble books in the proper order but were meaningless to the reader. Today, PHP code embedded in Web pages functions likewise, by marking the locations where texts will be assembled, and by directing operations and execution. Contemporary imperatives for action are not passed between printers and binders, but between Web browsers and the programs running on Web servers. Because it is the markup output of languages such as PHP that is passed to the browser, not the code itself, PHP creates a ghost in the machine, disrupting the ability of "View Source" to reveal what lies beneath the rendering of the source code of a Web page. (Notably, Ajax extends this change even further by disrupting the static nature of the viewable source, regardless of its genesis in executables or files.) This shift raises the bar for Web readers who would be doers. Given the increasing irrelevance of "View Source," the need for appreciating what Lev Manovich calls "database logic" grows as dynamic, database-driven Web sites become more common on the Web.

In "From Cyberspaces to Cyberplaces: <img>, Narrative, and the Psychology of Place," Rudy McDaniel and Sae Lynne Schatz use seminal theorists of image and text such as Roland Barthes and W. J. T. Mitchell to map a transformation regarding how we consider an increasingly image-oriented Web. For McDaniel and Schatz, the narratological strength of <img> is the most important characteristic of the Web's visuality. Images can provide a richness of meaning and facilitate the construction of communities because of the shared narratives common to images of all kinds (photographs, typographic elements, even the backgrounds and iconic elements used to brand many Web sites). It is now less useful to think of a single cyberspace, a concept established in the text-only spaces of Gopher, Telnet, and the early Web. Instead, following Mark Nunes, the Web should be imagined as a collection of cyberplaces, taking into account important differences in theorizing place as opposed to space. McDaniel and Schatz conclude with analyses of specific images by using four concepts from environmental psychology—imageability, legibility, wayfinding, and image schemata. In turn, they demonstrate the value of these concepts for theorizing the Web's usage of image.

In "<table>ing the Grid," Bradley Dilger plots the history of <table>, showing that its influence and use extends far beyond the intended use prescribed by the W3C. Graphic designers unhappy with HTML's limited design control appropriated <table> to create gridlike design systems patterned after the grids of modernist graphic design. The combination of tables for layout, along with pushes for graphical simplicity that are reinforced by the Web's limited typography, empowered the emergence of a design aesthetic similar to the international typographic style. Dilger shows how, in turn, this aesthetic's clean lines and orthogonal grids have been transferred to CSS, which use a box model and allow positioning of elements. Notably, the Web's embrace of the international typographic style extends beyond graphic design to its other values: efficiency, minimalism, and order. Dilger concludes by discussing implications for control raised by the inflexibility of Web-based grids.

Finally, Cynthia Haynes's afterword eschews the usual summary and synthesis common to this genre of writing. Recalling the motivation of Burke's casuistic stretching we make above, Haynes repeats his demand for continual attention to casuistry. Haynes's focus is the `<meta>` tag, which provides metadata to a Web page. She points out that the Hollerith machines used by Nazi Germany to track millions of victims of the Holocaust did so by simple conventions disturbingly similar to contemporary metadata standards. "If tagging is encoded at the very intersection of 'international business' *solutions* and political expediency," she writes, "then keypunch operators implemented (by proxy) the original `<meta>` tag, and sorting machines became the search engines ranking and optimizing the fate of millions of Jews." Haynes writes a prehistory of `<meta>` that, fittingly, concludes *From A to <A>* with some of the most profound and difficult questions asked by any of our contributors. Synthesizing the work of Lev Manovich with Steven Katz, she asks how we might balance questions of narrative—that is, our stories, our lives, our human richness—with their machine representations. In this way, Haynes situates her own essay as a casuistic meta tag for those that precede hers. Haynes's argument is a warning for our contributors, ourselves as readers of this collection, and ourselves as users of the Web. Markup is by no means an innocent practice, institutionalized or performed by individuals. In the age of markup, we must continue to build vocabularies of meaning so that we may better understand our rhetorical and writing practices in relationship to cultural, institutional, and ideological environments.

Works Cited

Burke, Kenneth. *Attitudes toward History*. Boston: Beacon Press, 1961.

Fuller, Matthew. *Behind the Blip: Essays on the Culture of Software*. Brooklyn: Autonomedia, 2003.

Fuller, Matthew, ed. *Software Studies/A Lexicon*. Cambridge, Mass.: MIT Press, 2008.

Heilker, Paul, and Peter Vandenberg. *Keywords in Composition Studies*. Portsmouth, N.H.: Boynton/Cook Publishers, 1996.

Kirschenbaum, Matthew G. "Virtuality and VRML: Software Studies after Manovich." *Electronic Book Review*. 29 August 2003. Accessed 5 October 2004. http://www.electronicbookreview.com/.

Liu, Alan. *The Laws of Cool: Knowledge Work and the Culture of Information*. Chicago: University of Chicago Press, 2004.

McLuhan, Marshall. *The Gutenberg Galaxy.* Toronto: University of Toronto Press, 1962.

Nelson, Carey, and Stephen Watt. *Academic Keywords: A Devil's Dictionary for Higher Education.* New York: Routledge, 1999.

Nelson, Ted. *Geeks Bearing Gifts: Version 1.1.* Sausalito, Calif.: Mindful Press, 2008.

Ong, Walter. *Orality and Literacy: The Technologizing of the Word.* New York: Routledge, 1982.

Stivale, Charles, ed. *Gilles Deleuze: Key Concepts.* Montreal: McGill-Queen's Press, 2005.

Swiss, Thomas, ed. *Unspun: Key Concepts for Understanding the World Wide Web.* New York: New York University Press, 2001.

Ulmer, Gregory L. *Electronic Monuments.* Minneapolis: University of Minnesota Press, 2005.

———. *Heuretics: The Logic of Invention.* Baltimore: Johns Hopkins University Press, 1994.

Williams, Raymond. *Keywords: A Vocabulary of Culture and Society.* New York: Oxford University Press, 1983.

TARRYING WITH THE \<head\>

The Emergence of Control through Protocol

THOMAS RICKERT

\<head\> is distinguished from the \<body\> container element. Often the first element to appear in a Web document, \<head\> can contain a document title, metadata such as the author's name or page keywords, links to scripts, style sheet information, and other code. \<head\> is a container tag; it does not perform an operation, affect the appearance of a page, or alter layout. Before current search technology that uses algorithms to index the Web, search engines located Web pages by the meta content included in \<head\> tags. Lycos, Netscape, Alltheweb, and other early search engines scanned keywords listed in the \<head\> section in order to match queries. Although such keywords still play a role in search, \<head\>'s usage of keywords is limited. A Google search, for instance, does not limit itself to the keywords located within the \<head\> tags but instead relies on other methods to produce page rank results.

\<forehead\>

HTML's \<head\> element is surprisingly well named. In its basic structure and function, it shares an affinity with the human head. This is borne out even at the metaphoric level, at which the tags for \<head\> and \<body\> are considered separate, albeit connected. And in terms of coding, the \<head\>'s functions affect but do not appear directly in the \<body\> content. Of course, such a separation already contains within it a priori assumptions (Cartesian, to be precise) about human existence. But all this suggests that there is more than bare functionality at work in choosing these metaphors and in achieving the specific coding form HTML takes. We are well accustomed to understanding that fiction, film, and poetry, from the most serious of enterprises to the most blatant of entertainment scams, is replete with material from the cultural imaginary. Why should code be any different? If it lends itself by associations of word and structure to the human head and/or mind, perhaps even richer forms of affinity are readily available.

But a first warning: the cultural studies–dominated 1990s overloaded us with the idea that anything in the world can be read as a text—film, culture, politics, movements. Such work granted us much insight. At the same time, the hermeneutic form of analysis tended to transfer its textualizing to the object itself: if punk rock could be read as a text, then there was something textual about the way punk style manifested itself in the lifeworld, connecting music with iconography with fashion with attitude. In delving into the associative logic connecting the head and the `<head>`, I want to move us toward seeing how it grasps us more powerfully than interpretative hermeneutics typically allows us to see something beyond the interpretive, readerly/writerly link. An understanding of code in terms of social impact or ideologies of technology, although certainly available and potentially of substantial interest, also implicitly reduces its material and immaterial agency. Code is also "code" for the oblique bluntness of immanence, which will here mean, among other things, control, protocol, and material persuasion. Thus, association as it is deployed here does more than define the new connective logics of networks; it takes part in them in such a manner that it opens up conflicts concerning intent and manifestation, design and implementation, dispersion and centralization, freedom and control that also emerge with and help forge networks. We might say that the head and `<head>` both mirror and take part in structures, a permutation of which is a constitutive split (or fissuring). However, this is no return to structuralism. We shall see that if structures do not march in the streets, nevertheless a certain efficacy remains as they patrol networks in the form of protocol. And yet protocol breeds its counterforces, an idea I will explore at the end with an examination of plagiarism.

\<headjob\>

The `<head>` element is one of several that compose an HTML document, and as I have suggested, it complements the hierarchal arrangement of content. Significantly, the `<head>` is placed ahead—or, given that code is vertically as well as horizontally arranged, it is placed atop the `<body>`. The `<head>` element sets off the title and other information that is not displayed on the Web page itself. In other words, the `<head>` element contains metainformation about the document that is not, strictly speaking, the content that appears throughout the `<body>`. `<head>` elements are information about information, data about data. A good deal can be stuffed into the `<head>`, including the `<link>`, `<meta>`, `<style>`, and `<title>` tags. With the `<meta>` tag, for example, one can include such metainformation as document titles and keywords for use by search engines. In this way, data talks to data, with none being shown in the Web page itself. An exception would be the title, which can appear as the caption of a document's browser window. The other tags can be used to define a number of other functions, such as JavaScript pop-up windows and pull-down menus, redirecting specific browsers to skip certain pages or go to alternative sites, and so on. In these and other cases, the `<head>` elements initiate actions that do not appear directly as content on the Web page.

Of course, each of these uses of the `<head>` tag is more involved than my quick gloss indicates. This is especially the case for tags that have a representational function, specifically tags that supply content about the first page. For instance, search engines will seek descriptive information about a page in the `<head>`, then use that data to retrieve and display a search result. Thus, keywords and titles contribute to the hierarchical organization of links that correspond to the search terms entered by a user. One would presume that the keywords used actually correlate with the content of the Web page, but this is not necessarily so. The `<head>` can contain keywords and titles of whatever sort, despite the convention that some sort of accuracy or honesty prevails (what we might call, as we will see later, a weak protocol). However, it is also the case that not all search engines parse the `<title>` keywords for their searches—Google, for example, does not. In such a case, the `<head>` element is a descriptive representation of the Web page content, but one of limited utility for actually finding, sorting, and ranking such content. For example, if we search for a Web site—say, Ebay—we type "ebay" into the search engine, and voilà, in 0.11 seconds, we get a list of Web sites to click on. In this case, given Ebay's popularity, we will not be surprised to see that it is listed in the first grouping of 10 out of 502,000,000 hits. The very first link is directly to Ebay.com and includes minimal information; right below it is a slightly expanded version of the first link, which includes sublinks to specific store categories: Motors, Electronics, Sporting Goods, and more. We see the title, "eBay—New & used electronics, cars, apparel, collectibles . . . " and a description, "Buy and sell electronics, cars, clothing, apparel, collectibles, sporting goods, digital cameras, and everything else on eBay, the world's online . . . " Google formats this content for ease of presentation and legibility, including the ellipses that truncate the lengthy descriptions. Given that the `<head>` elements can be manipulated by page designers so that particular search terms can be endlessly repeated and thereby skew searches that make use of such terms (i.e., front-loading an antipornography Web site with terms such as "sex," "XXX," "celebrity nudes," and "upskirt"), it makes sense that some search engines would weight differently the `<title>` and `<meta>` tags for finding, ranking, and evaluating content.

From a historical perspective, it seems perfectly obvious that something like the `<head>` element would be necessary. Nevertheless, it took time for this necessity to manifest itself. Tim Berners-Lee has remarked that early on he saw the need for escaping from the content of the document. Nevertheless, in the first published version of HTML (which mirrored trends already established by SGML, from which HTML is derived), the `<head>` was lacking, although tags such as `<title>`, which were to be placed in the `<head>`, did exist.[1] By the time of HTML 2.0 from September of 1995, however, the `<head>` was firmly established as a distinct section of an HTML document, clearly differentiated from the content of the `<body>`. Simplifying somewhat, we might then describe an HTML document as an entity with two main elements, the `<head>` and the `<body>`. It is ridiculously easy to point out that at a basic metaphorical level, a mind/body dichotomy is at work; and because the contents of the `<head>` are in fact excluded from the `<body>` content, the dichotomy is more than simply metaphorical; it is functional too. The `<head>` precedes the `<body>` and controls various aspects of it in a manner that is uncannily similar to humanist, modernist narratives concerning the controlling role of the head (and thought, soul, and spirit). The question is, what are we to make of this happy tangling of tendrils? What more is gathered together and put into play by ostensibly innocuous metaphors devised to label strictly functional coding decisions?

<headshrink>

I am claiming that it is more than a curious coincidence that the `<head>` tag takes the name it does. Certainly, we might describe our own heads in similar ways in respect to the body—that which controls but need not be seen in the way the body makes its display. But beyond this commonplace of everydayness, we start running deeper when we consider the matter from the perspective of psychoanalysis, which has always understood the human subject as being constitutively fissured between the mind and the body and, as we will get to, that which is conscious and that which is unconscious. But if we begin first with the prepsychoanalytic tradition, the head is "the engine of reason, the seat of the soul" searching for knowledge and meaning. More importantly, it is the traditional site of self-consciousness as that always-ongoing internal dialogue, sensation, and awareness that we equate with the physical body and the person. This would include all our mental contents, capacities, sensations, memories, desires, and so forth. The importance of the head as location and primary metaphor for the brain and thinking cannot be underestimated. The mind is the home of the I—more than home: Descartes' modernist project of ascertaining a single point of indubitable certainty came down to self-consciousness. The *cogito, ergo sum* is ultimately a headjob. Existence can be verified (logically, rationally) beyond a naive realism because one thinks and can be unmistakably aware of such thinking. Descartes connects this existential factum to the Ultimate Certainty that is God, and thereby re-constructs a Rational and Indubitable World. The head remains, as Daniel Dennett puts it, "the body's boss, the pilot of the ship" (77).

As is well known, what was held to be certain on Descartes' part has become considerably less so in the following years. Nietzsche, Heidegger, Freud, Lacan, and Derrida, from philosophical and psychoanalytic per-spectives, and Antonio Damasio, Steven Pinker, and Daniel Dennett (to name but a few) from scientific perspectives, have all taken Descartes to task. Although their perspectives differ, even radically so, they have in com-mon the unseating of pure apperception (à la Derridean "presence") and reason as the key foundations for human subjectivity. Slavoj Žižek, com-ing out of the psychoanalytic/philosophical tradition, puts it this way: "I am conscious of myself only insofar as I am out of reach to myself qua the real kernel of my being" (*Tarrying* 15). Žižek illustrates his point with an anecdote from the film *Blade Runner.* The detective, Deckard (Harrison Ford), discovers that Rachael (Sean Young) is in actuality a replicant, or artificial human, with implanted memories of childhood and family life. Rachael had thought she was human. Deckard is astonished. "How can it not know what it is?" he asks. Žižek suggests that Kant long ago supplied an answer to Deckard's existential question. From the Kantian perspec-tive, "the very notion of self-consciousness implies the subject's self-decenterment" (*Tarrying* 15). We see here a kind of radicalization and reversal of Descartes. Whereas for Descartes subjectivity and connected-ness to a world depend on one's perception of oneself as a thinking sub-ject, for Kant, Žižek argues, self-consciousness involves the opening up of a space one cannot know—consciousness is a twofold gesture, only one of which can be grasped by the self.

What is key here is that there is no conjoinment of the head and its experiential manifestation—its contents, if you will. Not only can we not see what goes on in the head, but even portions of our own heads remain opaque or closed to us. There is nothing mystical about this, though it can strike us as uncanny. Even in our own heads, we are not at home *(unheimlich)*. The thinker thinking is not at home, just like, as Barrett J. Mandel puts it, "the writer writing is not at home" (370). I understand that as I write and am in the flow. For example, right now, as I type this sentence, the words emerge through mental processes to which I have no direct access; further, the physical act of getting them typed into the computer is equally murky as I simply do it, without any apprehension of what is really going on. This does not mean that I could not look it up in a science text, but even then, whatever the level of understanding of synapses, neurotransmitters, nerve impulses, and so on, such an understanding will nevertheless remain for me, as we might say in Hegelese, for itself, never in itself. I might equally well trace all the strands and influences that work in and through me—books, conversations, emotions, people—but in the end, these are not equivalent to what I produce, which remains emergent. I can never take on the conscious experience of the intricacies of such cognitive emergence. Who I am, who we are, as beings, precludes such experiential understanding. In this way, we are fissured. Indeed, this was a key theme in Heinrich von Kleist's famous 1810 essay on grace and the marionette theater. Grace appears, as with an actor in performance, or any artful task really, to the extent that we are not self-awarely attentive upon the active processes themselves. Hence the singular import of marionettes: as wooden manufactures in performance, they are merely responses to string pulling, and their perfect maneuvers will far exceed the capabilities of any human dancer, who would be hampered not only by strength and skill but self-awareness too, which can humble even the greatest of artistes.

All these, the ideas of self-consciousness as impediment to perfection, are perhaps well understood. Every artist, athlete, and public performer has felt such pressures, felt how they entangle one's efforts. But we can continue our fine-tuning, moving toward a general statement on subjectivity. For what is less understood even if it is so commonly experienced is the split sense of self. Subjects are always doubled in the head: we have ourselves as our own object. Contemporary thought has addressed this head split in several ways: Lacanian psychoanalysis refers to the split subject, one that is constitutively fissured; Derrida refers to autoaffection (see, for an early example, *Speech* 78–80; also see Clough); and other poststructuralist theorists have their own conceptions. What is significant here is that this fissuring is quite different from the mind/body duality introduced by Descartes. Note, however, that from these perspectives, the fissuring of the subject is not a problem to be surmounted, but rather the necessary condition for a more complex and nuanced understanding of what it means to be human. In the case of Lacan and Žižek, such a fissure motors our subjectivity, generating the *energia* driving our will to see and understand ourselves through others. The sense and need for community, in short, stems from our inability to be a self-sufficient, present-to-self node.

\<post->

This kind of thinking, sometimes collected (rather loosely) under the rubric of poststructuralism, highlights a shift from the thinking we tend to call modernist, which more or less adhered to a notion of the subject as an autonomous unity, albeit a complicated and conflicted one.[2] Still, Kant's Enlightenment defining goal of achieving independence, social autonomy, and the capacity for self-derived forms of critical judgment remained crucial for education, administration, and governance, coming to define the modernist citizen-subject (see "An Answer"). Indeed, such a rationalist framework came to define not only the individual, but also the individual within the state, which was also increasingly rationalized. As the head to the unruly body, so the monarch to the nation.[3] The head tops a truncated, hierarchized, tightly controlled, and limited sociopolitical network, epitomizing the rationalist organization it simultaneously propagates, administers, and polices.

What we see with the advent of poststructuralist thought is the breakdown and/or reworking of this model, and, to greater and lesser degrees, the legitimation of the nation-state it informs. Still, it is no surprise that nearly all cultural and political discourses would be embroiled in debates concerning this breakdown of formerly secured (or at least securable) lines of sociopolitical power and the constitutive dichotomies on which they were grounded. As we will see, this breakdown is actually just code for the emergence of newer modalities of control built on more robust network logics (which will include new players such as transnational corporations and global special interest organizations). Although technology has always been implicated in social organization and politics, both practically and theoretically, it does so in a fresh manner today, challenging various modernist tenets and ushering in tremendous material transformation in culture and politics. As Patricia Ticineto Clough argues, teletechnology in particular has contributed to the discourses concerning "the deconstruction of the opposition of nature and technology, the human and the machine, the virtual and real, the living and the inert, thereby giving thought over to the ontologization of agencies other than human agency" (2). In the technological age, we have many "heads"—a proliferation of agencies all complexly entwined. This is one of the lessons we derive from those who have begun theorizing an emerging posthumanist perspective (such as Clough, Donna Haraway, N. Katherine Hayles, Edwin Hutchins, Rodney Brooks, Andy Clark, and Kevin Kelly). Thus, part of what posthumanism refers to is the unseating of the autonomous human head as the engine of reason, the seat of the soul, along with a concomitant series of challenges and dispersions on questions of what it means to be human, to organize ourselves in communities, and to engage technology.

Decentering, dispersion, and connectivity become important aspects of postmodernism and posthumanism. They condense complex discussions about the way social organization emerges and changes and to what extent human being is itself technologized. As Michael Hardt and Antonio Negri point out, "Just as modernization did in a previous era, postmodernization or informatization today marks a new mode of becoming human" (289). Although such debates are spread across the social and political spectrum, and thus remain mostly beyond the concern of this chapter, I want to emphasize that they do underpin the development of the World Wide Web itself. Tim Berners-Lee, widely considered to be the inventor of the Web, acknowledges that "inventing the World Wide Web involved my growing realization that there was a power in arranging ideas in an unconstrained, weblike way." He goes on to explain that this realization was not a linear solving of a problem, but an accretive process derived from a swirling set of influences, ideas, and challenges (3). Berners-Lee suggests that his immediate environs, the CERN laboratory, was crucial, as well as a series of previous inventions and developments, including TCP/IP, SGML, hypertext, and the Internet itself.[4] Also of importance was the impetus for the Web's generation: the ability to link through a common form all the disparate forms of information, not only at CERN, but throughout the world; the ability "to create a space in which everything could be linked to anything" (Berners-Lee 4). As he puts it, his task was to put two existing things—the Internet and hypertext—together (6).

We now get to the essence of the matter, at least for early and still significant conceptions about the World Wide Web. The vision Berners-Lee had was one that encompassed "the decentralized, organic growth of ideas, technology, and society" with "anything being potentially connected with anything" (1). The Web was to be unrestricted and "out of control" (99). This vision is heady and grandiose. It suggests underlying conceptions of communication, community, politics, ethics, and more. In particular, it advocates the vertiginous power of connectivity and the dispersion of centralized control into (what appears as) a rhizomelike (if not lifelike) network. In other words, it takes a stand against the head as a central, organizing authority and the hierarchical structures through which it has traditionally operated. Certainly, his vision, by its very inventiveness and productivity, suggests real limits to the scope and utility of hierarchies, centers, and control in advancing human knowledge, technology, and innovation (1–2, 16, 207–9). We also need to recall that this vision informed the endlessly optimistic, romanticized views of early Web theorists, who tied the burgeoning Web to notions of radical democracy, hyperconnectivity, identity switching and play, alternative communities, freedom of information and access, and immaterial evolution. And even if the backlash against such overly naive views started early, these views are still with us. There is a freedom, of which the World Wide Web is the epitome, that comes with unseating the head and scattering it across a network of greater and smaller nodes—what might be characterized as a form of rebodying. Such dispersion is tied in the cultural imaginary to visions of newly empowered citizenry. Indeed, it marks a profound shift in our understanding of power. But how to grasp this shift?

\<protocol\>

I have so far discussed the \<head\> tag in terms of a variety of themes, which I will now bring together. First, I discussed the tag's function in HTML, noting that it functioned analogously to various psychoanalytic and poststructuralist models of human subjectivity. The \<head\> introduces a fissure into HTML's content, acting like an unconscious (unseen) agent, providing information, orchestrating search results, and so on. In effect, I am exploring the possibility that the structure of HTML in part mirrors the structure of human subjectivity, and that such a structure is more than just a happy accident. Žižek argues that there is a sense in which "human intelligence itself operates like a computer," and he will later argue that the logic of cyberspace already mirrors that of the social symbolic, meaning that there was always already something virtual about the symbolic (*Tarrying* 42; *Plague* 143). Although I will not be making that same argument, I want to take cues from Žižek and suggest that thinking about the head as a controlling element—in terms of metaphor, operations, and power—gives rise to particular insights about the Web that in turn have relevance for human interaction and rhetorical engagement.

We can see this when we turn to a consideration of what our vision of the Web is like, especially in its germinal form. Berners-Lee imagined the Web as ushering in an era of total connectivity and new freedom for sharing and furthering knowledge, and indeed, the dominant understanding of the Web is as a decentralized, rhizomatic, vital network offering unprecedented (though far short of absolute) forms of communicative sharing. Intriguingly, for Berners-Lee and Robert Cailliau, the Web was initially imagined to be even more interactive and less spectatorial than it has in fact turned out to be. The browser Berners-Lee originally designed on his NeXT computer was both a viewer and an editor, and he continually tried to sell browser developers on the idea of making their browsers editors too (Berners-Lee 44–45, 70–71; see also Petrie). He met with little success, however, which perhaps indicates that more was going on than technical complications in browser development, as was often claimed, but instead an early leeriness about connectivity in general. To put this point differently, I am suggesting that from the very start, the lure (and threat?) of radical freedom (chaos?), decentralization, and dispersion on the Web led to the establishment of controls as built-in design features and as emergent structural features. When I say emergent, I mean that such structural features were not necessarily deliberate from a planning sense, and that they emerged out of a necessary logic that parallels aspects of human existence. We can say, then, that it is no accident that the <head> is split off from the content; there is a necessity at work that prescribes the emergence of such a split.

We can characterize such structural features as protocols. Alexander R. Galloway, in his study of the material structure of the World Wide Web and the Internet, states that a "protocol is a distributed management system that allows control to exist within a heterogeneous material milieu" (8). Protocols are simple rules governing the various operations involved in computing. Berners-Lee describes them as a "few, basic, common rules . . .

that would allow one computer to talk to another, in such a way that when all computers everywhere did it, the system would thrive, not break down," and he lists the ones most important to the Web, in decreasing order of importance, as URLs (uniform resource locators, which he prefers to call uniform resource identifiers), HTTP (Hypertext Transfer Protocol), and HTML (Hypertext Markup Language) (36). As Berners-Lee emphasizes, protocols are the common rules that make it possible for the Web and the Internet to function without any centralized command and control. In that sense, protocols are absolutely vital.

Nevertheless, protocol has material consequences for how the Web functions, a fact that even Berners-Lee acknowledges when he calls the DNS system the "one centralized Achilles' heel by which [the Web] can all be brought down or controlled" (126). Galloway, expanding on Berners-Lee's discussion, argues that the Internet, far from being the untamable rhizomatic beast it is so often portrayed as, is in actuality a product of two machinic systems that are fundamentally contradictory, DNS and TCP/IP. The TCP/IP system is a series of protocols for transmitting data over the Internet; they ensure that the data are fragmented, sent via packet switching, then reassembled properly and completely. These protocols are radically distributed and autonomous; as Galloway points out, one technical manual even uses the word *anarchic* in describing IP (8). DNS, however, opposes such decentralization by recreating rigid hierarchies. Galloway describes DNS as a large database that "maps network addresses to network names" (8–9; cf. Berners-Lee 126–28). In other words, the dozen or so root servers that host the databases organize nearly all Web traffic into hierarchical, treelike structures. Galloway notes that "each branch of the tree holds absolute control over everything below it," and he cites several examples in which a Web site was shut down, despite protests, by the communications company who owned the service provider (9–10). Although it may be unthinkable, it would be entirely possible to remove entire countries by modification of the information contained in the handful of root servers (10).

<control>

It would be alarmist to pursue such a line of thinking too far. It is highly dubious that a country will be removed from the Internet, and the shutting down of Web sites through the acts of corporate agents further up the hierarchical tree is likely to remain uncommon. The larger significance is not simply that it could happen, but that control exists despite radical decentralization. Furthermore, it is equally significant that such control emerges precisely through connectivity itself. Protocols do not exist solely in their own domain (or plane). Rather, protocols are nestled within each other, and in this too, the head has a part to play. For example, consider what we call "the Internet." It actually consists of Link, Internet, Transport, and Application protocols, arranged in layers (though it should be noted that there are other ways of conceptualizing these layers than the schema I am following here). Each layer finds itself encapsulated within the immediately preceding layer; for example, the Application layer is encapsulated within the Transport layer, and so on (Galloway 39). Thus, an HTML document, which exists on the Application layer, has a direct (nested) protocological relationship with HTTP, which is on the Transport layer. The HTTP protocol begins by parsing the HTML document, then creates a "special HTTP *header* that derives from the original object [the HTML document]" and describes it in various ways, then attaches this header to the beginning of the HTML document, like this example from Galloway (51):

```
HTTP/1.1 200 OK
Date: Sun, 28 Jan 2001 20:51:58 GMT
Server: Apache/1.3.12 (Unix)
Connection: Close
Content-Type: text/html
```

In this way, head elements provide the means for different layers of protocol to interact with each other to accomplish tasks. The header redescribes the original object for the next object; in this case, the HTML object is redescribed in the header for the HTTP object. Another way of putting it is that the larger protocol (HTTP) simply rewrites the smaller protocol in a manner fitting for the larger protocol (Galloway 51). Each protocol layer adds its own, new headers, rewriting the previous layer's object for its own protocol, while also making possible the connections between the various layers that enable the completion of tasks.

From a variety of standpoints, including the social and technological, the articulation of various protocols with each other, as by the nesting of head elements, is a marvel. Nevertheless, to consider the matter with a more associative logic, as we have been doing so far, tells us that this marvel has its dangers. This is not simply a matter of sounding a warning so that we can generate critique, either. It is not as if such protocols can be reversed, nor am I making any such argument. Thus, we can hold such advancements to be marvels without immediately taking some critical position on the matter. Nevertheless, more is at stake here. Although we might hold that it is important to recognize that the Internet, despite its common portrayal as something decentralized, rhizomatic, and uniquely positioned to offer radically new freedoms, is in fact highly controlled, this recognition is of limited usefulness. What is of far greater import is the kind of control that emerges. In a nutshell, this examination of the `<head>` demonstrates that control has not in fact been dispersed, despite the dissolution of many of the centers, hierarchies, and command structures characteristic of our understanding modernist, or disciplinary, society. Further, such control is enabled by fissures that are bridged by phenomena such as headers, which rewrite or otherwise transform one element for the utility of another. Thus, I argue, in the posthuman age, the head comes to mean something else than it formerly did: instead of central control, it means protocol. And protocol is sundered from the plane of content, as the emergence of thoughts and actions is sundered from direct, conscious apprehension. There are thus two key splits: `<head>` from `<body>`, and of greater import today, the TCP/IP–DNS fissure of the `<head>` itself.

Of clear interest to rhetorical theory is a consideration of how such ideas apply to the social realm. One way to address this is to compare protocol with a key modernist text, Kant's programmatic essay on enlightenment. Kant, like many modernist thinkers, propounded enlightenment as the means for an ongoing project of freedom, of liberation from superstition (such as religion) and authority. Enlightenment is emergence from self-imposed immaturity (41). Though it is likely to happen only slowly, it was held to be within anyone's reach. Indeed, immaturity is from Kant's perspective self-imposed, suggesting that becoming enlightened is an onus upon the individual. Still, enlightenment is not an absolute, and it cannot lead to total freedom. Kant's argument is quite subtle on this matter. He writes:

> Nothing is required for this enlightenment, however, except *freedom;* and the freedom in question is the least harmful of all, namely, the freedom to use reason *publicly* in all matters. But on all sides I hear: "*Do not argue!*" The officer says, "Do not argue, drill!" The taxman says, "Do not argue, pay!" The pastor says, "Do not argue, believe!" (Only one ruler in the world says, "*Argue as much as you want and about what you want, but obey!*) (42)

Officer, taxman, pastor—already a social trinity! And if it corresponds well with Foucault's diagram of (institutional) power in disciplinary society, so much the better. More importantly, tucked away in the parenthetical state-ment is Kant's mildly indirect reference to Frederick II (the Great) of Prus-sia, who, himself enlightened as Kant would have it (45), articulates where control resides when direct authority becomes subject to the endless, nameless inquisitiveness of enlightened rationality. One is free to argue, so long as one obeys (and if one is not so reasoning, one is stuck in lazy immaturity and cowardice). Here we see a fundamental contradiction of the Enlightenment. On the one hand, one is given by Kant a radical injunc-tion: argue, use your reason freely, and argue about what you will. Yet this potent freedom is tempered by a counterinjunction, obey!, that frustrates reason's power. For Kant, Law marks a limit for the freedom that can be achieved through reason. Law may well be subject to rational debate, but in terms of social functionality, it marks enlightenment's limit. Arguably every aspect of law can be rationalized, save one, adherence itself, as if soci-ety has never much advanced beyond the "Because I said so!" to the child's "Why?" Kant's *"Obey!"* thus accomplishes significantly greater rhetorical work than it at might first appear. Kant's early counter-Enlightenment critic Johann Georg Hamann in turn pointed out the reality of Frederick II's army, which Kant conveniently ignores: "With what kind of conscience can a reasoner & speculator by the stove in a nightcap [i.e., Kant as a lazy, armchair philosopher] accuse the immature ones of *cowardice,* when their blind guardian has a large, well-disciplined army to guarantee his infalli-bility and orthodoxy?" (147).

Although Berners-Lee may not have a statement as pithy as Kant's, the two can be fruitfully compared in order to see the subtle differences in the social and material diagram of control operational in the posthuman world. Connect, send, organize, build, say, access anything you want, but obey—protocol. Punning off Aleister Crowley, we could put it as "do what thou wilt, in accordance with protocol." As Berners-Lee points out through-out his book, achieving global standards is difficult, but they are precisely what enables the Internet (36, 188). For example, Berners-Lee discusses how at the first World Wide Web Consortium meeting in December 1994, a major topic of conversation was the fear that the extension of HTML might lead to its fragmentation, which was "seen as a huge threat to the entire community" (98). The proposed solution, however, is telling. Rather than impose a standard, they drew up a recommendation. In other words, they set up a protocol for the development of HTML. Naturally, such a pro-tocol, being more social, is looser than a material, network protocol. But in a sense, this is precisely my point. The new grid of posthuman control is no longer disciplinary, but protocological. I am making a case that it is of significance for rhetoric that we are now in the era of the protocol and recommendation. This means for us today, just as enlightenment meant for Kant and others then, that we have a newfound freedom. Unlike the Enlightenment paradigm, we have shifted from the regime of obeyance to one of recommendation. For the Enlightenment, the head still maintains command, though it may not be central or purely authoritarian. Yet it is the stand-in for Law. In this, our posthuman age, we say instead, "You are free to say, think, do what you like, so like as it falls within our minimum protocols and meets our few recommendations." Failure to comply is risky. The Law still functions, of course, so one still faces the threats of disciplinary society—prison, institutionalization, rehabilitation, and so on. It is not as if disciplinary society has disappeared. But increasingly, control works in ways that have nothing to do with hierarchical, centered heads.

As Hardt and Negri argue, contemporary "control society" (a term they derive from Deleuze) is a new form of inducement without regimentation or normalization (344).[5] For example, one is free to max out one's credit card, make late payments, run up fees, and within certain limits and penalties, still manage to function socially. Similarly, Nikolas Rose observes that a "politics of conduct enwraps each individual . . . in a web of incitements, rewards, current sanctions and of forebodings of future sanctions which serve to enjoin citizen to maintain particular types of control over their conduct" (246). Such protocols, so loosely written into the everyday fabric of life, seem to allow a great deal of freedom—freedom, perhaps, to pick whatever route one wants. But as Galloway points out, "protocol makes one instantly aware of the best route—and why wouldn't one want to follow it?" (244). And there's the rub. Not to follow credit's protocols produces traces—headers—that show up in yet other protocological systems. The late payments are rewritten into a data system that models the credit card holder for the credit card company; this data system is reported to a company that assigns a credit risk number; this number is combined with other data to create the portrait of the consumer by a credit reporting agency; when the consumer seeks to buy or rent something, that credit report is conveyed to the decision makers. In each layer, protocols are in operation, and by means of headers or their equivalents, these protocols are nested together.

The idea of protocol is of great significance for rhetoric, for it involves an entire style of management, control, persuasion, and communication, one that bridges material and social realms. Protocol, like the `<head>` tag, functions in the background, away from the direct arena of content; it orchestrates behind the scenes, or as a material element of the scene. Protocol, in a word, is the new, nonhierarchical manifestation of the head; it is thoroughly material, yet not equivalent to (the plane of) content. It is a model of how control emerges precisely where it seems to have evacuated.

As a way of making these associations and ruminations more concrete, I would like to conclude with a brief consideration of plagiarism. Certainly, the taking of another's work is a topic of heated discussion in the public realm. Authors, musicians, academics, politicians, journalists, and more, both famous and not, are reputedly engaged in extensive plagiarism. Students, particularly in college, plagiarize rampantly, with an extensive cottage industry to support their efforts if material found on the Web is not enough. But plagiarism involves more than this rather simple take on it suggests. Charges of plagiarism are potential threats for many new digital art forms, particularly song mash-ups, trailer remixes, and so on. Further, by extension, it can be said to be representative of an entire cultural shift, where digital technologies allow the endless reproduction and distribution of content, including images, music, film, video, print, code, and software. In this sense, plagiarism can stand for what is blacklisted, fairly or not, as piracy and theft as it covers the entire digital spectrum. Indeed, the same students who are lifting content off the World Wide Web are also copying CDs for each other and downloading files from the impossibly massive online inventory of digitized material.

What is odd about all this is the sharp discrepancy between those who decry such practices and those who find them an average if convenient part of the contemporary everyday. So clearly there is an ethical issue involved, not just the shrill yelping of profit-minded corporations or overworked teachers. Yet students appear strangely unmoved. We should be mindful here, for their moral ground tone is likely a greater indicator of the future than any legal and moral quick fixes offered by their elders. Certainly, students know that in terms of the social code, it is wrong to plagiarize. Yet unlike in the past, when honor systems seemed to have much more bite, today's students seem strikingly blasé about it. If they are caught, they seem to be more upset about breaking the Eleventh

Commandment ("Thou shalt not get caught") than about the actual plagiarism or piracy.

Why is this? One answer might be that today's students simply lack the moral fortitude of past students, but this seems unlikely. One could easily pick any number of issues about which today's students demonstrate exemplary moral acumen. A hint about what is occurring may come to us if we consider plagiarism not as a moral violation—which is to say, the formation of a bad subject in accordance with the precepts of Foucauldian disciplinary society—but as a declination to follow protocol. As we have seen, protocol is not just an isolated phenomenon but rather the diagram of control writ large for the digital world. And as I have argued, protocol works less like Law than as a form of second nature. Instead of a sign announcing "One Way—Do Not Back Up" and the threat of a ticket, it is a spiked grate puncturing your tires if you do so. That is, the commonsensical, enabling aspects of protocol tend to preclude alternatives. Indeed, part of the success of protocol is that it seems all to the good. Sure, one can take another direction, but why would one want to? Why buck things when they seem to making such excellent headway? For this reason, among others, resistance to protocol is extremely difficult. Unless, of course, the protocol is itself weak—that is, of limited purchase or scope. Thus, we might see most forms of plagiarism and piracy as so enabled by digital technologies that protocol has less bite than it might.

Galloway states that it is only in the context of protocol that we can understand computer hacking. Hacking is mostly understood, he argues, in terms of contemporary liberalism, by means of issues such as private property, financial gain, and access restriction (157). Galloway argues that this picture of hackers is of limited usefulness, and that what hackers are really about is working within the logics of code and protocol. Given that code and protocol are often synonymous, we can see that hackers are ultimately drawn to the possibility of exploiting protocol and making it work otherwise. Galloway argues, "What hacking reveals, then, is not that systems are secure or insecure, or that data wants to be free or proprietary, but that with protocol comes the exciting new ability to leverage possibility and action through code" (172).

I am suggesting that we might more profitably understand plagiarism/piracy through the lens of protocol rather than that of liberalism (or, put differently, across different head diagrams). The reason contemporary students so seldom feel guilt or remorse about plagiarism or file sharing (unless they are caught, of course) is that for them, it is not an issue of the ownership of ideas and property. This simply does not move them. Like hackers, today's plagiarist is out to leverage the system, to exploit the bugs in the university's (and by extension society's) educational and publishing codes; the music/video pirate avails himself or herself of a thriving industry of software, networks, and content, fully ensconced at this point and in no real danger of disappearing despite the significant attention of industry and government to curtail it. Like hacking, plagiarism and piracy are counterforces to protocol, but ones that work by means of the very logics of protocol.

Plagiarism, then, is better understood as a way of exploiting bugs already in the system. We could paraphrase the hacker Acid Phreak, who, by obtaining a few scraps of information, demonstrated how much real data, including credit information, he could find on John Perry Barlow; more outrageously, Acid Phreak shifted the blame from himself to Barlow, asking, "Now, who is to blame? ME for getting it or YOU for being such an idiot?!" (qtd. in Galloway 168). Perhaps these are harsh words, but tellingly, they are in accordance with precepts of protocol. Plagiarism is little different, I suspect, for many of our students. Are they to blame when the information is out there, so readily available, and so much is riding on obtaining good grades and degrees—when so much is at stake for them, whether they wanted to be in college or not? Piracy too: regardless of your, the reader's, personal take on the matter, the premium corporate entities place on retaining full rights to and profits from content facilitates the switch to grassroots digital invention, cooperation, sharing, and production—particularly when the digitally networked infrastructure is so nourishing and its contents so ripe.[6]

My intent has not been to conclude with a defense of plagiarism or piracy.[7] My intent has been to show that the (modernist) head discourse of liberalism may provide less insight than we would like into understanding how and why plagiarism/piracy occurs with such frequency today, and with such lack of conscience. By concluding with a discussion of plagiarism as an issue better understood in terms of protocol, I have opened the discussion to a consideration of the give and take of control in an era of hyperconnectivity and the ideological beckoning toward new openness and freedom. We see that networks and network architectures are political, and that politics is increasingly aligned via protocol. And to come full circle, we have seen that the <head> tag is a part of the ongoing extension of protocol as the contemporary diagram of control in a posthuman world. Just as it suggests a way of understanding human subjectivity as fissured, with much of our mental activities opaque to consciousness, so too the `<head>` operates behind the scenes of Web content. Taking cues from Žižek, I have argued that there are further parallels between the logics of cyberspace and the social symbolic, and I conclude with the idea that the `<head>` helps tag the emergence of protocol as a (posthuman) form of control in a nonhiearchical, decentralized world.

Notes

1. A very early version of HTML, dating from 1992, can be found at the World Wide Web Consortium (W3C) Web site, at http://www.w3.org/History/19921103-hypertext/hypertext/WWW/MarkUp/MarkUp.html. It is useful to compare this rudimentary version (derived from SGML) to HTML 2.0 and later versions. The amount and speed of growth is enlightening.

2. It should be pointed out that Žižek claims that ultimately Lacan is not a poststructuralist. See *Sublime* 153–55.

3. Steven Toulmin's *Cosmopolis* provides a good overview of this narrative concerning modernity.

4. Robert Cailliau, one of Berners-Lee's collaborators, also claims that environment and attitude—a kind of culture, as he puts it—are crucial to the inventive, developmental process. He attributes the emergence of crazily inventive approaches to computing in part to late 1960s California "flower power" culture. "Pure hackers don't wear business suits," Cailliau asserts. "The guys that come up with the ideas don't" (Petrie).

5. I will note without exploring further that Hardt and Negri state that control operates through three primary regimes: the bomb, money, and ether (345). It should be clear from my discussion that Internet protocol is very much a part of what they rather metaphorically refer to as ether. For more on control society, see Deleuze.

6. This perspective can be captured by punning off Bertold Brecht's famous quote on robbing a bank: "What's the crime of pirating a label's music compared to starting a new label?"

7. Nevertheless, art forms such as the music mash-up or movie trailer remixes, for all their promise, are not protected under current copyright law, except insofar as various loopholes (exploits) can be found, such as defining the work as parody. Conditions will undoubtedly change as law continues to evolve, but nevertheless, the world, I think, would be poorer without works such as Danger Mouse (*The Grey Album*, a mash-up of Jay-Z's *The Black Album* and the Beatles' *The White Album*), *Brokeback to the Mountain* (a trailer remix of *Brokeback Mountain* and *Back to the Future*), and so on, ad infinitum. I suspect that these and other such works will in the future be accorded more importance and status than currently.

Works Cited

Berners-Lee, Tim. *Weaving the Web: The Original Design and Ultimate Destiny of the World Wide Web*. New York: HarperBusiness, 1999.

Berners-Lee, Tim, and Dan Connolly. *Hypertext Markup Language—2.0*. 22 September 1995. Accessed 14 January 2009. http://www.w3.org/.

Clough, Patricia Ticineto. *Autoaffection: Unconscious Thought in the Age of Teletechnology*. Minneapolis: University of Minnesota Press, 2000.

Deleuze, Gilles. "Postscript on the Societies of Control." *October* 59 (1992): 3–7.

Derrida, Jacques. *Speech and Phenomena: And Other Essays on Husserl's Theory of Signs*. Translated by David B. Allison. Evanston, Ill.: Northwestern University Press, 1973.

Dennett, Daniel. *Kinds of Mind: Toward an Understanding of Consciousness*. New York: Basic Books, 1996.

Galloway, Alexander R. *Protocol: How Control Exists after Decentralization*. Cambridge, Mass.: MIT Press, 2004.

Hamann, Johann Georg. "Letter to Christian Jacob Kraus." Translated by Garrett Green. In *What is Enlightenment? Eighteenth-Century Answers and Twentieth-Century Questions*, edited by James Schmidt, 145–53. Berkeley: University of California Press, 1996.

Hardt, Michael, and Antonio Negri. *Empire*. Cambridge, Mass.: Harvard University Press, 2000.

Kant, Immanuel. "An Answer to the Question: What Is Enlightenment?" In *Perpetual Peace and Other Essays*, translated by Ted Humphrey, 41–48. Indianapolis: Hackett, 1983.

Kleist, Heinrich von. "On the Marionette Theater." 1810. Translated by Idris Parry. Accessed 15 January 2009. http://www.southerncrossreview.org/9/kleist.htm.

Mandel, Barrett J. "The Writer Writing Is Not at Home." *College Composition and Communication* 31 (December 1980): 370–77.

Petrie, Charles. "Robert Cailliau on the WWW Proposal: 'How It Really Happened.'" *Internet Computing Online* 2, no. 1 (January/February 1998). Accessed 13 December 2004. http://www.computer.org/.

Rose, Nikolas. *Powers of Freedom: Reframing Political Thought*. Cambridge: Cambridge University Press, 1999.

Toulmin, Stephen. *Cosmopolis: The Hidden Agenda of Modernity*. Chicago: University of Chicago Press, 1990.

Žižek, Slavoj. *The Plague of Fantasies*. New York: Verso, 1997.

———. *The Sublime Object of Ideology*. New York: Verso, 1989.

———. *Tarrying with the Negative: Kant, Hegel, and the Critique of Ideology*. Durham, N.C.: Duke University Press, 1993.

2 `<b> </b>`

Exploring Rhetorical Convergences in Transmedia Writing

SARAH J. ARROYO

"Be bold!" So goes a memorable Wikipedia editing guideline: be bold in editing pages. But the boldness Wikipedia seeks differs from that offered by the `<b>` tag, which makes text appear in boldface type. Like `<i>`, `<s>`, and `<u>` (italic, strikethrough, and underline), `<b>` is a nonsemantic tag: it provides visual information, as opposed to describing the structure or purpose behind an annotation. Advocates of standards-compliant Web design suggest `<strong>` as an alternative to `<b>`, and software such as WordPress follows suit, inserting `<strong>` tags when editors click the "bold" button in the WordPress text editor. Is moderated strength, then, a more palatable form of boldness? Indeed, the Wikipedia guideline, quoting Edmund Spenser, suggests so: "Be bold, be bold, and everywhere be bold," but "Be not too bold."

> It's the old dream of rationalism. Build a comprehensive system and drive the ambiguity out of it. Minimize the miscellaneous. Turn language into a machine and our machines will work wonders. People are too fallible. It is a dream that has failed over and over, not for a lack of trying.
>
> **DAVID WEINBERGER,** *EVERYTHING IS MISCELLANEOUS*

DAVID WEINBERGER'S COMMENTS encapsulate the goal of this entry for the bold tag: `<b> </b>`. `<b> </b>`, a simple typographical tag, brings to the forefront a juxtaposition of issues ranging from ancient rhetoric to rhetorics for new media in order to illuminate the importance of writing in and for multiple media platforms. Boldness, as we will see, highlights some of rhetoric's central areas of contention because it brings several rhetorical divisions—historically set up in opposition—to the forefront and creates the conditions for intermingling these divisions, giving a broader and more encapsulating role for writing in digital culture. The notion of a single narrative written by a single person is no longer accurate, and traditional separations such as virtual/material, rhetorical/physical, surface/depth, and visual/structural converge to contribute to the growing rhetorics for new media. The goals for this entry are lofty indeed; yet at the core is <b> </b>: the tag that will open up possibilities for rhetoric and writing in digital culture. I begin by defining <b> </b> in order to show that it is almost always pinned against its counterpart, <strong> </strong>. This binary creates the underlying tension that will intertwine the aforementioned similar rhetorical divisions. I then explore each division at the threshold of its convergence to render a rhetoric of potentiality for digital design.

What Is It to <b> </b>?

<b> </b> serves first as an exemplar for how mere appearance carries much significance, despite its negative historical connotation (similar to mere rhetoric). This notion, while serving as a defining feature, will also be the first riff for the tag because I will connect its devalued status to devalued forms of rhetoric. <b> </b> is considered a physical tag because it simply changes the look of the text to bold. Appearance and concern with how something looks are commonly viewed as shallow and nonintellectual. Accordingly, most HTML instructional guides caution against using <b> </b>, because the tag does not provide any hints about semantic meaning; again, it simply changes the text to bold for the sake of appearance. The following represents a sampling of HTML guides, which will highlight <b> </b>'s devalued status and will begin to show parallels between the language of markup and rhetoric. We will see <b> </b> set up in direct opposition to its logical counterpart, <strong> </strong>, which is the preferable tag for highlighting text.

HTML Dog, a Web site that encourages best practices of markup, places <b> </b> in the category of presentational, thereby ascribing to it simple ornamentation. Dan Tobias devotes an entire page of his site, Dan's Web Tips, to explaining the vast differences between physical and logical tags and defines bold as a "physical (or visual) HTML tag as compared to STRONG," which is defined as a "logical (or structural) tag. Physical tags represent specific, visual effects that are intended to be reproduced in a precise manner and carry no connotation as to their semantic meaning" (Tobias). Similar to the function of ornamentation, visual effects simply remain on the surface and are not concerned with articulating any (deep) meaning.

So far, we already have myriad oppositions with which to contend: bold/strong, physical/logical, visual/structural, surface/depth, ornamentation/semantic meaning. The left half of each division is the unfavorable, as one might predict. Tobias continues in his explanation advocating logical tags over physical ones: "The advantage of using the logical rather than the physical tags . . . is that your meaning will be more precisely conveyed." Thus, the use of logical tags allows software programs to render the meaning of the text, and the appearance may not be as the designer intended on all platforms and media devices. In fact, Tobias suggests that "hard line purists may say to use logical tags at all times, while some graphical designers advocate using only physical tags (because the logical ones have the tendency, distressing to those of the visual mindset, to be rendered in widely varying ways)." This tip suggests that those of the visual mind-set are not open to display variances—something I will explore later because display variances are nearly unavoidable, the result of the wide range of viewing capabilities available when online.

Eric Tilton's "Composing Good HTML" is perhaps the most value-ridden guide. It stresses: "The elements of HTML describe *what* your information is, not *how* it should be displayed. All it involves is a bit of trust. The trust you must have can be summarized by the following rule: *If* you mark up a document so that your information is labeled as what it is instead of as how it should be displayed; *Then* browsers will render it in a way that is appropriate and professional looking." Although these instructions are certainly meant for all tags (not just <b> </b>), they bring forth the main point of contention for boldness. Boldness, with all its semantic associations set aside for the moment, is seen as rigid, nonconforming, and physical; "being bold" is not only a negative attribute, but also one devoid of any meaning. Ironically, then, "b" is not about the ontological question of "to be" at all; <strong> </strong> carries the weight of being and defining "what is." The above trust factor is worth further elaboration; if we define an object's content, then we are more certain to have our intended meaning realized. Intended meaning, then, is much more important than surface-level appearance. We are thus cautioned against using the physical tags altogether because they leave no room for interpretation and simply act as commands to render appearance.

These guides assume the superiority of defining objects over providing visual direction, thereby recreating the split found in literate culture that values print over the pictorial. Thus, my essay aims to complicate the simple divisions created by the two styles of tagging by exploring markup writing as conceptual design. Conceptual design focuses on appearance as well as functionality, so what something is does not overpower how it works on a given device or media platform. As the Immersive 360 design team puts it on their blog: "A new breed of designer is emerging who can flow freely between the real and the virtual; between media and across cultures" (Admin). This new breed of designer is a writer, multimedia producer, videographer, and participant in digital culture. He or she does not create things by sticking to the old divisions of virtual/real, logical/physical, surface/depth, and visual/structural. Instead, these divisions converge and intersect in ways not possible in the nondigital world, opening up innumerable creative possibilities. The following sections will explore three different areas of rhetorical convergence, offering three explanations for how conceptual design works in digital culture. These convergences are not meant to synthesize the binary oppositions. Rather, I hope to show how the oppositions intermingle to generate new possibilities. We will first travel back to ancient rhetoric to confront the historical division between rhetorical learning and physical training, which will bring us toward the discussion of surface-level knowledge and depth models of mastery, bringing us to the exploration of the visual and structural. Finally, I will discuss how the idea of conceptual design works better for digital composition in multiple media platforms.

The Virtual Material World: The Convergence of the Rhetorical and Physical

In "Bodily Pedagogies: Rhetoric, Athletics, and the Sophists' Three Rs," Debra Hawhee revisits the gymnasia of Athens in ancient Greece to show the inexorable connection between rhetorical and athletic training. She returns particularly to the sophists, because they typically did not stay in one place to interact with students. Hawhee points out that "most—if not all—sophists passed through the city's gymnasia at some point" (144), and turns to Isocrates to argue that "the sophists offered a distinctive approach to rhetorical pedagogy derived from physical trainers" (145). We then are directed to a crucial difference between the pedagogy of the sophists and typical learning situations: "Rather than focusing on material learned—the sophists didn't have a curriculum in the modern sense of a 'subject matter' to be 'covered'—sophistic pedagogy emphasized the materiality of learning the corporeal acquisition of rhetorical movements through rhythm, repetition, and response" (160). These three Rs, as Hawhee calls them, represent the convergence of rhetorical learning and the bodily movements associated with each rhetorical canon. The cadence involved in repeating rhythmic movements, for instance, cannot be separated from the content learned. Hawhee goes on to suggest that "from this spatial intermingling of practices there emerged a curious syncretism between athletics and rhetoric, a particular crossover in pedagogical practices and learning styles that contributed to the development of rhetoric as a *bodily art*: an art learned, practiced and performed by and with the body as well as the mind" (144). Hawhee's claims are extremely important for this entry because she quite convincingly merges the training of the mind and the body.

Hawhee is not the first to emphasize the convergence of mind and body. In a review of scholarship on material rhetoric, "Weighty Words: The Materiality of Rhetoric," Catherine Hobbs writes: "The constant effort to nullify the body and its demands paradoxically returns our energetic attention to the body. And nowhere do we exercise more fine control, more mind–body integration and harmony, than in speech and writing" (116). The body, then, perpetually returns, demanding our attention and continually resisting deflection. D. Diane Davis might equate the body with what she terms the *exscribed*, which, she suggests, exemplifies that which has been deflected by speakers and writers who attempt to assert authority and control. However, the exscribed, no matter how tough the deflection, keeps coming back, haunting writers and their writing nonetheless: showing up to destroy the facade of clarity expected, for instance, when using semantic tags. Even if we assume that intentions are met and clear communication exchanges occur, we still have the exscribed "gesturing to us from the door," as Davis puts it (134). The exscribed, the physical, becomes a sort of prankster, always disrupting clear communications exchanges. We are thus constantly reminded of its presence—its boldness, so to speak—and to simply deny it is impossible.

Combined with the above descriptions of the constant return of the deflected, Hawhee's focus on the canon of rhetorical delivery provides the most applicable way to explain the convergence of the rhetorical/physical and virtual/material; she links rhetorical delivery to the convergence of rhetorical and athletic training (156). While obvious, she acknowledges, the canon of delivery has been most neglected in modern eras, simply because of the prevalence of logic in rhetorical training. Hawhee elaborates on the historical neglect of delivery here: "Perhaps one reason for delivery's oversight is its sheer corporeality, as well as its attention to the less rational qualities of rhetorical speeches such as volume, rhythm, and cadence" (157). She also reminds us of the necessary bodily strength involved in delivering a powerful oratory, thus highlighting the literal melding of the physical and logical. I suggest that much of this argument about the physical aspects of rhetoric can be transferred to the act of writing, and especially multimedia and transmedia writing.[1] For instance, for multimedia writers to design an effective piece, they must not only work with text, audio, and visuals, but they also must be acutely aware of timing, rhythm, cadence, repetition, interaction, and response—notions that would be familiar to the students Hawhee describes in the gymnasia of Athens. Accordingly, transmedia writers must anticipate the physical roles users will enact as they interact with the material. Thus, the physical aspects of digital design also blur the distinction between the physical off-line world and the virtual online world.[2]

I hope that Hawhee's description of rhetorical delivery as the corporeal acquisition of rhetorical movements resonates as I continue to riff on the connection to designing in new media. In *Small Tech: The Culture of Digital Tools*, Byron Hawk et al. describe the concept of small tech, which begins to blur the hard line division between the physical and virtual worlds. They write: "Small tech highlights the complexity of the threshold between the material world of big, physical things and the virtual worlds of conceptual, affective communication and calls for the attention of humanities scholarship" (xi). The complex threshold serves as an exemplar for the convergence of the material/virtual and physical/logical. These binaries have been engrained in educational institutions for centuries: what we know and can articulate in a rational manner carries much more weight than what we can only feel and sense. Knowledge residing in the body is simply irrational and emotional, and is discarded in favor of what can be articulated logically. However, the idea of the small tech threshold supports the conceptual, affective knowledge arising from the work of digital design. The participatory nature of digital culture allows for deep interaction among users that brings affective knowledge to the forefront. Digital design embodies in a virtual sense the canon of rhetorical delivery as described by Hawhee: Hawk et al. explain this morphing of the physical and virtual when they write, "If big, physical things have already met their potential, materialized as they are in reality, virtual things are potential realities" (xii). In other words, if rhetorical delivery in the material world entails the literal combination of mind and body, its embodiment in the virtual world would exist as the potential for what interactions may become.

Henry Jenkins also acknowledges the participatory nature of digital culture and its necessary affective dimension when he describes communities forming online. These communities are created through "voluntary, temporary, and tactical affiliations, reaffirmed through common intellectual enterprises and emotional investments" (*Convergence Culture* 27). Emotional investments are important here because these are the very things that bring digital communities together and sustain them. Further, the process of acquiring knowledge (built around emotional investments) keeps these communities "dynamic and participatory" (54), as Jenkins points out. Thus, the small tech threshold might be explained as virtual, potential bodily knowledge, which collapses the distance between the real and virtual. Hawk et al. also explain: "Increasingly miniaturized components and goods redefine the threshold between materiality and the conceptual, cognitive, and affective dimensions of every day life" (x), which directly attributes the convergence of the material, virtual, physical, and rhetorical to the different media capabilities with which people engage: Web browsers, PDAs, and other handheld devices.

Boldness as Bravado: The Convergence of Surface and Depth

So far we have seen definitions of <b> </b>, made connections to ancient sophistic practices of melding the rhetorical and physical, and taken these arguments into the virtual realm where affective knowledge and communication can be realized as a virtual embodiment of the canon of delivery as the sophists viewed it. We will now circle back to boldness itself to explore the convergence of surface (feigned knowledge) and depth (mastery). If boldness is deflected, if it is Davis's prankster, then perhaps the boldness carries bravado, feigning mastery and rigidness and continually disrupting the intent of communicating what something means by using semantic tags; yet ironically, sticking to a depth model of mastery, trying to capture the whole of a concept, is also illusory. I suggest that in electronic writing, rhetorical mastery is not as important as combining knowledge sources into a functional network. Writers (designers) compose spatially (in order to link from surface to surface) and not semantically. David Weinberger, in *Everything Is Miscellaneous*, laments that when we try to define and categorize in digital culture as we do in print culture, we "re-create the same problems faced by the traditional categorizations of knowledge: Human topics are too big and squishy to fit well into any one set of boxes" (193). Instead, when writing online, people collaborate with others to make a given concept come to life.

Jean-François Lyotard discusses the problem of striving for semantic mastery in "Resisting a Discourse of Mastery." Here Lyotard claims that an answer is only interesting insofar as it elicits a new question; an answer is not the assertion of a solution or mastery of a given concept, because a supposed solution closes off further inquiry. The goal, then, is continual questioning and linking to other concepts, which eventually creates a web or network of knowledge. This concept does not sit well in a world where we are encouraged to provide concrete answers to complex problems. For Lyotard, "to write is to allude to something else which is not easily communicated" (2). This allusion creates the conditions for additional linking; it does not purport to provide answers but encourages active inquiry. The job of making knowledge is to constantly invent and create by way of constructing multiple networks. This generative conception of knowledge runs counter to the notion from print culture that we must master a concept before we communicate its tenets to others.

However, the generative metaphor for knowledge does not simply favor surface-level, associative thinking over depth models of knowledge. Rather, it merges the surface-level necessity of linking with the depth requirement of expertise and specialty. I turn to the Web sites Innocentive: Where the World Innovates and the Insight Community to serve as exemplars for the convergence of surface-level and depth-based knowledge. Innocentive brings together solvers and seekers. Solvers are self-described experts in their fields, and seekers are corporations or public institutions that need particular problems solved. Seekers post their problems, and solvers— typically only tied together because of their investment in the problem at hand—collaborate to come up with potential solutions. Similarly, the Insight Community comes up with large challenges to global issues and asks users to vote on an issue to tackle. Once the issue is decided, users also work together to find solutions. The environmental organization Zero Footprint used Insight to come up with the zero footprint database, which, among other things, allows users to calculate their daily environmental footprint. One of their challenges, for example, came from American Express Corporation, which sponsored conversations "concerning how small businesses are dealing with the financial crisis" (https://www.insightcommunity.com/cases.php/). These sites use social networking to promote their causes while relying on combined, collective wisdom, not individual mastery, to wrestle with complex problems. A single solution is not the goal; rather, potential solutions to be implemented in varying contexts come out of these sites, thereby forming an ever-growing network of possibilities. The network (surface) merged with the expertise and emotional investment of the users (depth) creates a rich environment for innovation.

That is to say, the network is just as important as the expertise of the members: they are symbiotic in their relationship.

Recharacterizing the Image:
The Convergence of the Visual and the Structural

With the convergence of surfaces and depth models resonating, I turn briefly to Ron Burnett's *How Images Think* to help explain the convergence of the visual and structural in digital design. Recall that `<b> </b>` is a tag that only denotes appearance, whereas `<strong> </strong>` gives commands regarding the structure of an object. Again, we see boldness as weak and strongness as, well, strong. Images (and anything visual) are traditionally linked to mere appearance (one's image) or a replica of something that exists in the real world. They are seen as inferior to text and to interpreting meaning from text. As we have seen in the previous discussions, it is no longer possible to make such simplistic divisions, and digital culture has complicated not only the definition of the image, but also the role images play in everyday communication exchanges and relationship, identity, and community building. Burnett writes: "Images have been transformed into linguistic, emotive, and embodied instruments that guide individuals through their experiences of nature and culture" (82). This explanation—particularly the notion of embodied instruments—elevates the image to the small tech threshold discussed earlier, where the physical, rhetorical, real, and virtual collide. In other words, this description of the image does not allow for a simple division between any of those elements. If images guide us through our experiences, then they are crucial for building identities and relationships and for understanding particular cultural experiences. In fact, Burnett suggests, "Cyberspace . . . is a trope for a new kind of human interaction. . . . This affects . . . the role and impact of imagescapes on the process of communications" (82). But these interactions, particularly in social networking sites, are not without structure. Images also act as interfaces, and interfaces structure or influence our decisions while working in digital culture. Building a social networking site and participating in one both require a tremendous amount of linearity and structuring. On Ning, for example, designers must plan the site to maximize users' interactions. Otherwise, the site will fail. Because the coding is built in these sites, designers can customize what is already there or use the preloaded code to import widgets and other objects to make the site successful in appearance and functionality.

Designing a social networking site exemplifies the convergence of the visual and structural even more aptly when we apply Weinberger's concept of the "ecology of the implicit and explicit." Weinberger explains: "Computers deal only with what they've been told, not with what's been left unsaid. And that is causing a disruption of the delicate ecology of the implicit and explicit" (154). What is missing is the visual or physical element necessary for human interaction to be successful. Images escape definition because of their complexity and wide variety of possible meanings. Weinberger remarks, "Making complex, meaningful phenomena explicit can leave us rudderless, force us to oversimplify, and result in statements that are incomplete and misleading" (156). By linking Burnett's description of images in digital culture to Weinberger's articulation of the semantic objects' demand for being explicit, we can see that images are no longer on the underprivileged side of the binary. Instead, they have complicated the notions of logical definition and clear articulation. Weinberger continues: "The folksonomies that are emerging bottom up are characterized by ambiguity, multiple classification, and sort-of kind-of relationships" (196). "Sort-of kind-of" relationships define the convergence of the visual and structural because room for ambiguity exists alongside necessary structures. These relationships also support implicit meanings, which necessarily exist in any participatory site. In fact, "a seamless whole that drives out ambiguity would also drive out the richness of implicit meanings" (195). Accordingly, implicit meanings, combined with necessary explicit elements, create the best conditions for community building to occur.

Implicit meanings change depending on context and what is not explicitly stated by the code, for example. Implicit meanings, Weinberger suggests, are "fragile"; they are "easily lost as culture moves on" (195). These meanings come about because of shared knowledge. As I mentioned earlier, many communities form online around an emotional investment that cannot necessarily be defined. The participants of the community keep it functioning, dynamic, and beneficial; implicit meanings surface through the interactions of people and their collective understanding of the structural information provided by the sites or databases involved. We see this happening most often in sites providing database-driven suggestions when users respond collectively to the suggestions. These responses often catch the attention of the site designers, who can then go into the site and make necessary changes. The convergence of the explicit (structural) and implicit (visual and cultural) drives the users' experiences and provides meaningful exchanges among parties.

Crossing Platforms: Transmedia Writing and Conceptual Design

The three previous sections aimed to complicate binary divisions prevalent in rhetorical and HTML instruction. Through this exploration, I hope to have highlighted the importance of writers as conceptual designers and transmedia producers who create across multiple media platforms for a variety of viewing capabilities. The conceptual designer/transmedia writer is automatically collaborative and participatory and works at the center of all of the sites of convergence discussed in this essay. Content can be developed, changed, merged, and morphed, producing what Hawk et al. call "potential realities" (xii). These potential realities are the crux of participatory culture, where all participants believe their contributions matter. Hawk et al. also suggest that virtual objects are "nothing more than densely interwoven sets of relations" (xii), a statement that supports all of the riffs with which I have engaged in this essay, allows for varying user experience, and delicately balances implicit and explicit content. This entry is not meant to be a literal call for using <b> </b> or not using it. Instead, I hope these points of rhetorical convergence, all spawned by boldness, can contribute to the growing number of investigations into writing and participatory culture, and can help reconceptualize the rhetorical and cultural act of writing across media platforms into conceptual design.

Notes

1. Henry Jenkins explains that "a transmedia text does not simply disperse information: it provides a set of roles and goals which readers can assume as they enact aspects of the story through their everyday lives" ("Transmedia Storytelling 101"). This is an important part of mixing the rhetorical and physical because without physical enactment and participation from users, stories and other digital productions would not exist. That is, transmedia writing provides the conditions for users to participate and become collaborators in the stories themselves.

2. A resurgence of research on the canon of delivery has occurred in the wake of the study of Web 2.0 and participatory culture. The physical aspects of digital design and their connection to rhetorical delivery are explored at length by Collin Brooke, who recasts rhetorical delivery as performance, James Porter, who devises five components in a theoretical framework for digital delivery, and Jim Ridolfo and Danielle Nicole DeVoss, who link rhetorical delivery and velocity to strategic recomposition. These efforts address the canon of delivery from the explicit perspective of new media writing, but I believe Hawhee's pre–Web 2.0, overt concentration on melding the rhetorical and physical provides the best source of connection for this entry.

Works Cited

Admin. "5D: The Future of Immersive Design." *Immersive 360 Blog.* 1 October 2008. Accessed 8 January 2009. http://www.blog.immersive360.com/?p=18.

Brooke, Collin. *Lingua Fracta: Towards a Rhetoric of New Media.* Cresskill, N.J.: Hampton Press, 2009.

Burnett, Ron. *How Images Think.* Cambridge, Mass.: MIT Press, 2004.

Davis, D. Diane. "Finitude's Clamor, or Notes toward a Communitarian Literacy." *College Composition and Communication* 53, no. 1 (2001): 119–45.

Hawhee, Debra. "Bodily Pedagogies: Rhetoric, Athletics, and the Sophists' Three Rs." *College English* 65, no. 2 (2002): 142–62.

Hawk, Byron, David M. Reider, and Ollie Oviedo, eds. *Small Tech: The Culture of Digital Tools.* Minneapolis: University of Minnesota Press, 2008.

Hobbs, Catherine. "Review: Weighty Words: The Materiality of Rhetoric." *Rhetoric Society Quarterly* 31, no. 2 (2001): 116–22.

HTML Dog. "Presentational Tags." *HTML Dog.* Accessed 6 January 2009. http://htmldog.com/.

Innocentive: Where the World Innovates. Accessed 10 January 2009. http://www.innocentive.com/.

Insight Community: Expertise on Demand. Accessed 10 January 2009. http://www.insightcommunity.com.

Jenkins, Henry. *Convergence Culture: Where Old and New Media Collide.* New York: New York University Press, 2006.

———. "Transmedia Storytelling 101." *Confessions of an Aca/Fan,* 22 March 2007. Accessed 31 July 2010. http://www.henryjenkins.org/2007/03/transmedia_storytelling_101.html.

Lyotard, Jean-François. "Resisting a Discourse of Mastery: A Conversation with Jean-François Lyotard." Interview with Gary Olson. *JAC* 15 (1995): 391–410.

Ning: Create Your Own Social Network. Accessed 18 August 2008. http://www.ning.com.

Porter, James E. "Recovering Delivery for Digital Rhetoric." *Computers and Composition* 26, no. 2 (2009): 207–24.

Ridolfo, Jim, and Danielle Nicole DeVoss. "Composing for Recomposition: Rhetorical Velocity and Delivery." *Kairos* 13, no. 2 (2009). Accessed 10 March 2009. http://35.9.119.214/13.2/index.html.

Tilton, Eric. "Composing Good HTML." Accessed 7 January 2009. http://www.ology.org/tilt/cgh/.

Tobias, Dan. "Physical vs. Logical Markup." *Dan's Web Tips.* Accessed 5 January 2009. http://webtips.dan.info/.

Weinberger, David. *Everything Is Miscellaneous: The Power of the New Digital Disorder.* New York: Times Books, 2007.

3 alt

Accessible Web Design or Token Gesture?

COLLEEN A. REILLY

alt *is an attribute for* <img>*; it is primarily used to assist the visually impaired.
Assistive technology can read the text within* alt *attributes and thus give blind users
access to page content and function.* alt *provides the same service for those using
nongraphical browsers or who surf the Web with graphics turned off. When a reader
mouses over images,* alt *text appears as a tool tip. And finally, if for some reason, an
image does not appear,* alt *text alerts the reader of what was supposed to appear in
the space. In new media or net art, the* alt *attribute offers artists additional space to
leave commentary or to extend their Web projects. Jodi.org, an experimental Net art
site, famously displayed what appeared to be random blinking text. When the reader
clicked "View Source," however, the schematics for a hydrogen bomb were revealed
by use of an altered* alt *tag. The Web comic Achewood uses* alt *attributes to provide
a joke about the day's strip or metacommentary about characters interacting in the
strip. In these ways, alternative uses for* alt *complement its core function.*

> My text-based screen reader did not recognize most buttons or icons.
> Instead of showing me a picture, it just said "image." Instead of verbally
> indicating where a link would take me, it just said "link." If there was a series
> of pictures and linkable words or icons, my computer would simply say,
> "[Image] [Image] [Image] [Link] [Link] [Link]."
>
> **VALERIE LEWIS,** "[IMAGE] [IMAGE] [IMAGE] [LINK] [LINK] [LINK]: INACCESSIBLE
> WEB DESIGN FROM THE PERSPECTIVE OF A BLIND LIBRARIAN"

THIS CHAPTER EXPLORES alt="", or the alt attribute, used in HTML to
provide textual alternatives for visual content. alt has long been a marker
of the marginal, othered, and controversial on the Internet, starting with
the "alt.*" Usenet hierarchy, in which such discussion groups as alt.drugs,
alt.sex, and alt.rock-n-roll were created in the late 1980s (Bumgarner;

Lee; Sexton). Although the alt-ed spaces of the alt.* hierarchy and alt attributes in HTML were developed ostensibly to provide a means for participation and access for other voices and users, in practical terms, the resulting alternate Usenet and Web site experiences created separate yet unequal situations, leaving the original Usenet hierarchy and Web sites no more democratic or accessible than before.

This chapter argues that focus on alt attributes and other code-based accommodations is misplaced, leading too often to Web sites that are technically compliant but practically unusable. Concentrating on the minutiae of code often allows designers to absolve themselves from creating usable and satisfying experiences for all users, regardless of the means through which they access and interact with sites. That is not to say that alt attributes, inherently subordinate as image attributes in HTML (Slatin), can or should be ignored, but that the use of these and other code-based accommodation strategies needs to be part of a revisioned design process for maximum accessibility and usability (Pernice and Nielsen; Slatin and Rush). As this chapter demonstrates, numerous analyses of electronic resources reveal that the Web in general is far from being an inclusive and universally usable space (Davis, "Disenfranchising" and "Accessibility"; Hazard), just as our physical spaces often fall short of accessibility goals despite years of regulations, laws, and legal actions that resulted from the Americans with Disabilities Act. Truly revisioning the Web for maximum accessibility requires, in part, shifting our conception of disabilities from the medical model, concerned with assisting people with specific impairments with predetermined tasks by complying with a checklist of accommodations, to a social and civil rights model that situates all persons along a continuum of ability and highlights that marginalizing those less abled may mean someday marginalizing oneself (Burgdorf).

alt: A Way In for Other Voices?

References to alt in conjunction with the Internet may evoke the controversial newsgroup subhierarchies alt.sex.* and alt.drugs.*. Historically, the creation of the alt.* hierarchy stemmed from a dual desire to democratize the process for creating new newsgroups while simultaneously providing a space for undesirable topics that was distinguishable from the original seven (now eight) Usenet hierarchies. Although Usenet as a whole has been alt-ed by the visual Web and many predict its continued diminishment and eventual demise, in the 1980s and into the 1990s it was the central place for the online exchange of ideas and, later, files (Segan). The lore regarding alt.* reports that it humorously stood for "anarchists, lunatics, and terrorists" (Barr; "1987: The alt.* Hierarchy"); however, most users simply think of it as standing for "alternative," evocative of the other, the unsanctioned, and the illicit. alt.* is aligned with the alt attribute through its role in providing room, albeit nonmainstream and stigmatized, for marginalized users to participate in personally relevant conversations online, however objectionable they may seem. Like the alt.* hierarchy, the alt attribute facilitates a different experience of and interactions with visual Web resources, although the equality of this alternate representation is an open question (Hazard; Pernice and Nielsen).

The alt.* hierarchy was created in 1988 in reaction to the refusal of Usenet backbone administrators (often called the Backbone Cabal) to support newsgroups related to sex and drugs (Hardy; Lee 364). Before the creation of alt.*, a new Usenet group could only be created after a brief electronic discussion that yielded no serious objections among the news administrators who composed the Backbone Cabal and, after 1987, through a newsgroup-based voting process (Lee 365). The process has continued to evolve and now includes the submission of a thorough request for discussion (RFD) to the moderated group news.announce.newgroups, the discussion of the RFD by members of the Big-8 Management Board on news.groups.proposals, the optional polling of membership by the group's initiator or the Big-8 Management Board, to determine potential interest in the new group, and a vote of the Big-8 Management Board regarding whether or not to create the group ("How to Create a New Big-8 Newsgroup"). In contrast, from the start, alt.* groups subverted this centralized control, allowing individuals to create new groups by sending the appropriate control message to administrators who will carry the group (Barr; "How to Create an ALT Newsgroup"). Since its inception, alt.* has expanded exponentially to include thousands of groups related to specialized, bizarre, and potentially offensive subject areas (see a list at "Alternative Newsgroups").

One of the creators of alt.*, Brian Reid, recalls that its birth seemed to represent an embrace of openness and free speech on Usenet:

> The introduction of genuine free speech on the Usenet came with the development of the "alt." hierarchy, which now has more newsgroups than any other hierarchy. All other Usenet hierarchies require you to follow a definite procedure to create a new group in that category. However, a group in the

"alt." hierarchy can be created by anyone. The resulting set of newsgroups form a truly global democratic system, since each group in the "alt" category survives only if people show an interest in it. ("Alt Hierarchy History")

However, as several online information sites covering the creation of new alt.* groups indicate, the impression of this hierarchy—aptly termed—as an uncensored, fully democratic space free from centralized control is illusory (Barr; "How to Create an ALT Newsgroup"). According to "How to Create an ALT Newsgroup," although the rules for creating new groups are less arduous than in the big eight hierarchies, they still require the submission of a proposal for a new group to the alt.config group, whose volunteer cadre of news administrators ("gods") will issue a response to the proposal, which may consist mainly of outlining its flaws: "If they don't rip your proposal to shreds you can send a control message to get it newgrouped. There's nothing to stop you going ahead and newgrouping a group even if it receives a hostile reception, but the gods may well rmgroup it ('remove group') and it WILL NOT PROSPER." This process could be viewed as less democratic than the process required by the big eight hierarchies that at least allows for the polling of Usenet membership, although the degree to which such polls are conducted and/or influence the vote of the board is unclear ("Requests for Discussion (RFDs)").

Additionally, labeling a class of newsgroups alt.* has proved to be as stigmatizing as it is liberating. News administrators as well as Internet service providers can and have refused to carry the entire alt.* hierarchy, and cordoning off hosts of groups there actually makes this sort of wholesale censorship easier. For example, in June 2008, Sprint and Verizon stopped offering customers access to the alt.* hierarchy, retaining access only to the groups of the big eight, under pressure from the New York state attorney general, who claimed that Usenet sites on the alt.* hierarchy were propagating child pornography (McCullagh, "N.Y. Attorney General" and "Verizon"). Although only a "handful" of alt.* groups were found to contain any illegal content, the Internet service providers censored the entire hierarchy nationwide as a result of its stigmatized associations (McCullagh, "N.Y. Attorney General" and "Verizon"). Eliminating this hierarchy with, if examined closely, loads of perfectly legal but potentially inflammatory content proved far easier than closing off access to only specific groups accused of illegality.

In the end, any democratizing effects of alt.* also fail to make the remaining areas of Usenet more democratic or uncensored. The apparently more open space of the alt.* hierarchy has not influenced the main eight Usenet hierarchies to reduce the hurdles in the process for the creation of new groups. alt.*'s existence supports the impression that freedom of speech online has, at least in one space, been achieved, while in practice, many users find this freedom mythic or access to it eliminated by indiscriminate censorship. Similarly, as the discussion below demonstrates, the alt attribute and other code-based accommodations provide the appearance of inclusivity for differently abled users by permitting sites to be labeled as compliant through conformity with accessibility guidelines. However, these users' experiences of the sites often prove to be much less productive and satisfying than the site's rote compliance may promise.

Models of Disabilities, the alt Attribute, and Maximum Accessibility

The choice of "alt" for the attribute that provides text-based access to visual Web content is unfortunate because, as with the alt.* hierarchy on Usenet, it connotes an othered or marginalized track for reading that can be largely ignored by mainstream readers. Providing alternate text for content that some users cannot see without considering the whole of their experience and/or actually observing it does not result in an integrated site that provides maximum accessibility any more than relegating unpleasant or uncomfortable content to the alt.* hierarchy transforms all of Usenet into an uncensored and democratic space. This section examines social and integrative models of disabilities, discusses the appearance of alt in guidelines for and assessments of site accessibility, explains the connection of alt to the limiting medical and compliance models of disabilities, and suggests that designing for maximum accessibility requires shifting the focus, in light of the social model of disabilities, from the code to the creation of accessible and usable experiences for all users.

Medical, Social, and Integrative Models of Disabilities

Before the passage of the Americans with Disabilities Act (ADA) in 1990 and the subsequent 1998 revision of section 508 of the Vocational Rehabilitation Act, laws and public policies relating to people with disabilities were based on a medical model that focused on physical limitations and constructed people with identifiable impairments as patients (Riley 4–6). In contrast, the ADA focuses not on medical conditions but on the ways that the physical structure of public spaces needs to change to provide access. This shift away from the medical signifies a privileging of the social and civil rights models of disabilities (Riley 7). Tregaskis further clarifies the distinction between impairments and disabilities: "A distinction is drawn between the acknowledged inherent functional limitations caused to the individual by their impairment, and the additional socially created barriers caused by disability" (10). Thus, our physical and technological spaces are, in a sense, malformed by failing to provide usable interfaces for all potential users and readers.

The social model is productively extended to highlight the lack of a firm barrier between abled and disabled persons. As Burgdorf asserts, "Human beings do not really exist in two sharply distinct groups—those with disabilities and those without. The actual reality is what has been called a 'spectrum of abilities.'" This integrative model emphasizes that there are infinite degrees and manifestations of ability and that any person at any time, through accident or old age, may operate through a disability. As a result, in building design or information architecture, designating content for a particular group, such as the alt.* hierarchy, relegates uncomfortable content to a separate track; it does not acknowledge the potential movement from ability to disability and back that is possible for all readers and users.[1]

alt in HTML, Accessibility Guidelines/Analyses, and the Perpetuation of the Medical Model

In HTML, alt functions as an image attribute and appears in the code as follows: <img src="imagefilename" alt="alternate text that provides replacement content for the image/visual" />. References to alt attributes first appeared in the HTML 2.0 specification document in 1995 and make no mention of accessibility for people with disabilities, instead recommending their use to make content available for users whose access is hampered by technical factors, including lower bandwidth and slower connection speeds (Berners-Lee and Connolly). In developing their recommendations for HTML 4.0, the World Wide Web Consortium invited experts in Web accessibility to review drafts of the specification (Paciello), resulting in the inclusion of a separate section that outlines specific guidelines for the effective use of alt attributes in accommodating people with disabilities (Raggett, Le Hors, and Jacobs).

The proposed specifications for HTML 5 and the new Web Content Accessibility Guidelines (WCAG) 2.0 continue to highlight the use of the alt attribute and other code-based methods for providing textual alternatives for graphic elements, including the use of `<label>` elements to provide text labels for form controls and the `title` attribute to describe visual elements that are not easily described through other means (Hickson and Hyatt; Caldwell et al.).[2] Although the HTML 5 specification acknowledges an ongoing, online debate about recommending and/or requiring alt attributes (for two perspectives, see Gilman; Lemon), the specification still recommends their continued use as a central tool for accessibility and provides extensive guidelines for how to use them in the section on images (Hickson and Hyatt). Additionally, WCAG 2.0 covers text alternatives for nontext content in section 1.1 of its revised guidelines (Caldwell et al.) and explains the use of the alt attribute in the very first section, 1.1.1, of the site explaining how to meet WCAG 2.0 requirements (Vanderheiden et al.).

In research concerning Web accessibility, the appropriate use of alt attributes and text are identified as the most basic form of accommodation, often compared to curb cuts for people in wheelchairs (Comeaux and Schmetzke 458). Despite the increased coverage of the alt attribute in each new HTML specification, excerpts from analyses of a variety of sites representing numerous institutions and industries, from university libraries to the airline industry, demonstrate that the omission of alt attributes and meaningful alt text are repeatedly identified as both the central and most easily correctable accessibility problems (examples are in chronological order):

- Online brokerage sites: "Conversely, the Charles Schwab site used an image to convey the risk level of a particular mutual fund. The images were named *Risk 1* through *Risk 5*, and they had no ALT text associated with them. The image itself did not use a numeric risk level, but rather indicated a high, medium, or low risk level. It would have helped screen reader users to have the images named *High Risk, Low Risk,* etc., or to have the ALT text provide this information" (Pernice and Nielsen 46).
- News Web sites: Davis reports that regarding news Web sites, the percentage of home pages accessible to the visually impaired could be increased from 7 percent to 87 percent if alternate text were provided for all graphic elements ("Accessibility" 479).
- E-commerce sites: "Most of the barriers we encounter here are caused by missing ALT text—that is, image maps and buttons that lack the equivalent text alternatives that would make them accessible to

people using assistive technologies such as screen readers and talking Web browsers" (Slatin and Rush 17).

- Airline reservation sites: "The most commonly encountered fully supported error [in the 73 sites examined] was requiring alternate text on all images, with 84.9% of the sites having such error. Among the sites with such error, the average number of occurrences of this error was 39.7" (Gutierrez, Loucopoulos, and Reinsch 243).
- Library school and research library sites: "Researchers also found lack of alternative text to be the 'most frequently occurring barrier'" (Comeaux and Schmetzke 474).

Although each of these studies points to the ease of use for this accommodation, it is clear that alt attributes continue to be omitted and the Web continues to be inaccessible. Pernice and Nielsen estimate that "the Web is about three times easier to use for sighted users than it is for users who are blind or have low vision" (5).

The gap in usability can be linked in part to the influence of the medical model of disabilities that fosters a checklist mentality of accommodation. The medical model, as outlined above, assumes that there are easily identifiable disabilities affecting specific categories of people who can be accommodated by designs that meet the discrete requirements of guidelines such as WCAG 2.0. The checklist mentality perpetuates the impression that making a site accessible means checking it, often at the end of the design process, with an automated accessibility tool, such as Watchfire WebXACT (originally Bobby and now the IBM Rational Policy Tester Accessibility Edition software). Three design problems connected to the medical model of disabilities arise from this approach. First, the accessibility check is perceived to be necessary in order to make a site usable by an alternative group of disabled users, most of whom have obvious physical disabilities, such as visual impairment or motor issues.[3] Because these users are constructed as nonmainstream, not as the target users for the site, the site's failure to pass the accessibility check may not be seen as a critical problem; the site is still viewed as appropriate for the majority of users without obvious disabilities. Second, the checklist mentality implies that examining the site for accessibility is not an integral part of the design process; it is a separate process designed to accommodate literally alt-ed users. As a result, if the accessibility analysis is omitted from the design process for time or financial constraints, the omission is not seen as crucial. Finally, even if a site passes an automated accessibility analysis, many people may still find that site difficult or impossible to use. Completing the checklist only verifies that the required elements, such as **alt** attributes, are present; it says nothing about how effectively they are used (Gutierrez, Loucopoulos, and Reinsch). As with architectural accommodations, all of the requisite elements can be checked off, but they still might not work together to provide a satisfying or productive access to the building. For example, to be accessible, a building should have both a ramp and an automatic door button; but what if these accommodations were not at the same door? Or what if one, like the automatic door button, were missing? Visitors in wheelchairs who could use the ramp would still require assistance entering the building. On paper, according to the checklist, the building is accessible, or nearly so; yet in practice, it provides a poor experience. The prevalence of checklist mentality in Web design is exemplified in the analyses cited above that evidence a lack of adherence to the guidelines and a persistence of inaccessible designs.

In attempting to meet the letter of accessibility guidelines, some designers create the ultimate **alt** text: completely separate, text-only sites. As Hazard and others demonstrate, in many cases, these separate sites are incomplete or otherwise flawed. Separate text-only sites represent the most extreme application of the medical model where nonmainstream users are relegated to a completely separate space/interface. Hazard's study of research library Web sites in 2006 and 2007 found that twenty-one member libraries of the Association of Research Libraries provided text-only versions of their sites, approximately half of which were separate pages and half of which were dynamically generated (421). Most of the completely separate sites omitted important information, such as library contact information, hours of operation, and links to special collections, government documents, and employment opportunities (Hazard 424). Additionally, the separate pages were updated less frequently. The dynamically generated text-only sites were more accurate, complete, and timely, but were not completely accessible unless the parent site contained **alt** attributes and other code-based accessibility elements (426). Because text-only versions are generally inferior and creating them seems to absolve designers from making their graphic sites accessible (Pernice and Nielsen 55), designers are advised to avoid developing these separate and generally unequal, text-only sites and focus on designing sites that are accessible to all (35).

To adequately address accessibility, designers need to foreground fashioning a meaningful and inclusive experience for all users, regardless of their media of access and their level of ability (Pernice and Nielsen 9; Caldwell et al.). Creating such an experience necessitates moving beyond the code, the checklist, and strict applications of legal requirements. As Slatin and Rush argue, "The unfortunate aspect of reliance on government mandates comes from the sense that compliance—rather than maximum accessibility and usability—is the goal. . . . So, although the regulations have spurred interest in accessible design, we will see that to meet the spirit of the law, we must go beyond both the regulations and the law" (55). As discussed below, a less divisive, more social model of disabilities can potentially be more productive, calling for changing our spaces and interfaces, and integrating the experiences of all users into one design process and experience.

Maximum Accessibility and Usability

Designing for what Slatin and Rush term maximum accessibility requires revisioning the design process and incorporating all types of users, including those with a variety of disabilities, into the usability analysis process. The impetus to revision design draws on the social and integrated models of disabilities that encourage designers to reach users at all points on the spectrum of ability and, as Pernice and Nielsen explain, to avoid "view[ing] people with accessibility issues as a single big user group" (41). Maximally accessible design cannot ignore or dismiss code-based accommodations, such as alt attributes. In a reworked design process, the details of accessible code are viewed as necessary to make the HTML complete, but are supplemented by inclusive usability analysis and macro-level design techniques. As a result of constantly evolving design innovations, assistive technologies, and user needs, complete accessibility for all users is unlikely to be achieved in a static, absolute sense: "ensuring accessibility . . . becomes a matter of probability rather than certainty" (Spinuzzi 198). Attempting to maximize accessibility is nevertheless a laudable goal, essential for ethical design.

The number of Americans with "covered" disabilities equals roughly one-fifth of the population, or about 54 million people (Gutierrez, Loucopoulos, and Reinsch 240; Jaeger 519; Perez).[4] In outlining to whom the ADA applies, the law mentions both having an impairment and being perceived to have an impairment (Riley 7). As Bérubé argues, disabilities are one "area of social life in which the politics of recognition are inseparable from the politics of redistribution" (53). Many studies have demonstrated that people with disabilities of all sorts are economically disadvantaged by the lack of opportunities that result from inadequate access and diminished opportunities. Regarding the Web, Riley emphasizes, "The inaccessibility of the Internet is the most pressing civil rights problem facing people with disabilities today" (207).

Others note the potential, yet unfulfilled, of the Web for people with disabilities: "Most fundamentally, the Internet is liberating. For example, people with visual impairments can read the daily newspaper the minute it is published, rather than wait for a taped transcription" (Pernice and Nielsen 3). Individuals who have difficulty leaving home to meet with friends and family and accomplish daily tasks could use the Web to become more connected and independent (Pernice and Nielsen 17) and could make use of Web-based services that facilitate online banking, employment seeking (Loiacono and McCoy 399), and/or remote commuting (Russell).

As recent lawsuits such as *National Federation of the Blind (NFB) v. Target Corporation* have determined, simply meeting accessibility specifications is insufficient; the requirement is to provide a space in which all users can accomplish all tasks. In this class-action suit, the NFB claimed that Target violated the civil rights of blind users by denying them the ability to complete successful transactions on the Target site *(NFB Target Lawsuit)*. Neither the claim nor the terms of the settlement mention accessibility guidelines or code-based accommodations, such as adding alt attributes, but focus instead on restructuring the site to allow blind customers to effectively complete e-commerce transactions or apply for employment *(NFB Target Lawsuit)*. The focus is on providing these users with a satisfying and productive experience.

One important way to increase the likelihood that users of varying levels of abilities will have a successful experience with a site is to include individuals with a range of abilities into the usability analysis process (Correani, Leporini, and Paternò). As Cooper, Colwell, and Jelfs explain, people with disabilities view Web sites differently from typically abled users; they are more likely to be critical if the experience provided is inadequate and less likely to blame themselves (235). Usability analysis is a key element of accessible design, for as, Spinuzzi argues, users' continually shifting sets of available technologies and resources make "accessibility a moving target. . . . That is, if we examine accessibility as an object(ive) of user experience, we can only support probable configurations, not guarantee accessibility per se" (198). Usability analyses provide designers with a tool to identify potential impediments and revision designs.

Usability analyses incorporating people with disabilities have already recommended a number of revisions to the design process that may aid in producing more usable sites. Web designers are urged to consider the overall structure of information, including how it will be read from top to bottom by screen readers and the variety of routes users may take through the site, ensuring adequate means of identifying location and links and navigating site content. Others argue for designs incorporating semantic annotations that communicate to assistive technologies the function of HTML elements (Kouroupetroglou, Salampasis, and Manitsaris 274; Sevilla et al.). Last, for certain content, instead of using complex graphic designs, writers might turn to more inherently accessible forms, such as blogs and wikis.

The techniques for devising a revisioned, inclusive Web design process are as endless as the articles that propose and detail them. However, it is the social and inclusive model of disabilities that provides an important motivation for change. As many scholars emphasize, "improved accessibility for disabled users promotes usability for all" (Cooper, Colwell, and Jelfs 232; see also Hazard 420). As Slatin and Rush argue, "Taking the needs of people with disabilities as the starting point for design leads to aesthetically richer, more productive, and more satisfying Web experiences for *everyone*, not just for people with disabilities" (161). There are many examples of design features that help all users; "logical tab order on Web pages helps people using a screen reader, a screen magnifier, or an assistive technology related to motor skill challenges. It can also help power users who want to use keyboard commands," and "leaving white space between links on a Web page helps people with and without motor skill challenges hit targets accurately" (Pernice and Nielsen 40). The social and integrative models of disabilities make universally inclusive design imperative by emphasizing that at any time, any of us may find ourselves in another place on the spectrum of abilities, and none of us will wish to be constructed as alternative users, barred from the access we had heretofore enjoyed.

Conclusions

The **alt** attribute and other code-based techniques for use in accessible Web design cannot be completely disregarded; their use should be considered as a small part of changing design practices to facilitate maximum accessibility. Accomplishing inclusive design is quite difficult, however, and requires revising design practices to include a constant focus on the myriad numbers of devices and methods of access available to users of all abilities. If the literature is any indication, the Web is a long way from being a universally accessible space where all users can function and accomplish their goals. alt provides a paradox: it cannot be ignored, but a focus on alt attributes and other code-based minutiae as the basis for accessibility in design leads to missing the point and creating sites that provide separate and quite unequal experiences for many classes of users.

Notes

1. As Bérubé argues, unfortunately, the U.S. Supreme Court has not interpreted the ADA from this more inclusive perspective, because if it had, "more Americans would realize their potential stake in it" (54). Bérubé highlights three lower court cases on which the Supreme Court ruled in 1999 in which "people with 'easily correctable' disabilities—high blood pressure, nearsightedness—were denied employment. In three identical 7–2 decisions, the Court found that the plaintiffs had no basis for a suit under the ADA precisely *because* their disabilities were easily correctable" (54). In highlighting these cases, Bérubé emphasizes how the Supreme Court perpetuates the either/or model of disabilities, which focuses on a distinct, easily identified set of physical impairments, restricting the application of the law to this segment of the population and further marginalizing them in the process.

2. The details of the HTML and WCAG guidelines for the use of the alt attribute and other means for providing textual alternatives for graphic elements are beyond the scope of this chapter. However, these details can be accessed online in Hickson and Hyatt; and Caldwell et al.

3. Sevilla et al. point to the relative lack of concern for making Web resources accessible for users with cognitive disabilities, which often have fewer physical manifestations.

4. As Riley explains, numerous organizations that advocate for people with disabilities cite much larger numbers.

Works Cited

"Alternative Newsgroups." *Newsville: Virtual News.* 3 January 2009. Accessed 3 January 2009. http://www.newsville.com/.

"Alt Hierarchy History." *Living Internet.* Accessed 2 January 2009. http://www.livinginternet.com/.

Barr, David. "So You Want to Create an ALT Newsgroup." *Usenet FAQs.* 11 August 2001. Accessed 2 January 2009. http://www.faqs.org/.

Berners-Lee, Tim, and D. Connolly. "Hypertext Markup Language—2.0." *World Wide Web Consortium (W3C): Web Accessibility Initiative.* 1995. Accessed 3 January 2009. http://www.w3.org/.

Bérubé, Michael. "Citizenship and Disability." *Dissent* 50, no. 2 (2003): 52–57.

Bumgarner, Lee S. "The Great Renaming: 1985–1988." *Living Internet*. Accessed 5 August 2010. http://www.livinginternet.com/.

Burgdorf, Robert L. "Negative Media Portrayals of the ADA. The Americans with Disabilities Act Policy Brief Series: Righting the ADA." *National Council on Disability*. 20 February 2003. Accessed 2 January 2009. http://www.ncd.gov/.

Caldwell, Ben, Michael Cooper, Loretta Guarino Reid, and Gregg Vanderheiden, eds. "Web Content Accessibility Guidelines (WCAG) 2.0." *World Wide Web Consortium (W3C): Web Accessibility Initiative*. 11 December 2008. Accessed 18 December 2008. http://www.w3.org/.

Comeaux, David, and Axel Schmetzke. "Accessibility: Web Accessibility Trends in University Libraries and Library Schools." *Library Hi Tech* 25, no. 4 (2007): 457–77.

Cooper, Martyn, Chetz Colwell, and Anne Jelfs. "Embedding Accessibility and Usability: Considerations for E-Learning Research and Development Projects." *ALT-J: Research in Learning Technology* 15, no. 3 (2007): 231–45.

Correani, Francesco, Barbara Leporini, and Fabio Paternò. "Automatic Inspection-Based Support for Obtaining Usable Web Sites for Vision-Impaired Users." *Universal Access in the Information Society* 5 (2006): 82–95.

Davis, Joel J. "The Accessibility Divide: The Visually-Impaired and Access to Online News." *Journal of Broadcasting and Electronic Media* 47, no. 3 (2003): 474–81.

———. "Disenfranchising the Disabled: The Inaccessibility of Internet-based Health Information." *Journal of Health Communication* 7 (2002): 355–67.

Gilman, Al. "Re: [html4all] some reflections on @alt usage (and even sometimes @aria-labelledby . . .)." *W3C Public Mailing List Archives*. 28 April 2008. Accessed 5 December 2008. http://lists.w3.org/.

Gutierrez, Charletta F., Constantine Loucopoulos, and Roger W. Reinsch. "Disability-Accessibility of Airlines' Web Sites for U.S. Reservations Online." *Journal of Air Transport Management* 11 (2005): 239–47.

Hardy, Henry Edward. "The History of the Net." MA thesis. Grand Valley State University, Allendale, Mich., 1993. *Electronic Frontier Foundation*. Accessed 2 January 2009. http://w2.eff.org/.

Hazard, Brenda L. "Separate but Equal? A Comparison of Content on Library Web Pages and Their Text Versions." *Journal of Web Librarianship* 2, no. 2/3 (2008): 417–28.

Hickson, Ian, and David Hyatt, eds. "HTML 5: A Vocabulary and Associated APIs for HTML and XHTML." *World Wide Web Consortium (W3C): Web Accessibility Initiative*. 17 December 2008. Accessed 18 December 2008. http://www.w3.org/.

"How to Create an ALT Newsgroup." *Nylon.net*. 11 May 2008. Accessed 2 January 2009. http://nylon.net/alt/.

"How to Create a New Big-8 Newsgroup." *Big-8 Usenet*. 7 July 2010. Accessed 5 August 2010. http://www.big-8.org.

Jaeger, Paul T. "Beyond Section 508: The Spectrum of Legal Requirements for Accessible E-Government Web Sites in the United States." *Journal of Government Information* 30 (2004): 518–33.

Kouroupetroglou, Christos, Michail Salampasis, and Athanasios Manitsaris. "Browsing

Shortcuts as a Means to Improve Information Seeking of Blind People in the WWW." *Universal Access in the Information Society* 6 (2007): 273–83.

Lee, Hangwoo. "'No Artificial Death, Only Natural Death': The Dynamics of Centralization and Decentralization of Usenet Newsgroups." *Information Society* 18 (2002): 361–70.

Lemon, Gez. "HTML 5 Alternative Text, and Authoring Tools." *Juicy Studio.* 1 May 2008. Accessed 5 December 2008. http://juicystudio.com/.

Loiacono, Eleanor T., and Scott McCoy. "Website Accessibility: A Cross-Sector Comparison." *Universal Access in the Information Society* 4 (2006): 393–99.

McCullagh, Declan. "N.Y. Attorney General Forces ISPs to Curb Usenet Access." *CNET News.* 10 June 2008. Accessed 3 January 2009. http://news.cnet.com/.

———. "Verizon Offers Details of Usenet Deletion: alt.* Groups, Others Gone." *CNET News.* 12 June 2008. Accessed 3 January 2009. http://news.cnet.com/.

NFB Target Lawsuit Information Site. N.d. Accessed 5 December 2008. http://www.nfbtargetlawsuit.com/.

"1987: The alt.* Hierarchy." *Giganews Usenet History.* Accessed 2 January 2009. http://www.giganews.com/.

Paciello, Michael G. *Web Accessibility for People with Disabilities.* Lawrence, Mass.: CMP Books, 2000.

Perez, Stella. "Macromedia Accessibility Project (MAP)." *Macromedia.com.* 18 November 2002. Accessed 5 August 2010. http://download.macromedia.com/pub/accessibility/map_whitepaper.pdf.

Pernice, Kara, and Jakob Nielsen. "Beyond ALT Text: Making the Web Easy to Use for Users with Disabilities." *Nielsen Norman Group.* October 2001. Accessed 5 December 2008. http://www.nngroup.com/.

Raggett, Dave, Arnaud Le Hors, and Ian Jacobs, eds. "HTML 4.01 Specification." *World Wide Web Consortium (W3C): Web Accessibility Initiative.* 24 April 1998. Accessed 3 January 2009. http://www.w3.org/.

"Requests for Discussion (RFDs)." *Big-8 Usenet.* 8 July 2010. Accessed 5 August 2010. http://www.big-8.org/.

Riley, Charles A. *Disability and the Media: Prescriptions for Change.* Hanover, N.H.: University Press of New England, 2005.

Russell, Catherine. "Access to Technology for the Disabled: The Forgotten Legacy of Innovation?" *Information and Communication Technology Law* 12, no. 3 (2003): 237–46.

Segan, Sascha. "R.I.P. Usenet: 1980–2008." *PC Magazine.* 31 July 2008. Accessed 4 August 2010. http://www.pcmag.com/.

Sevilla, Javier, Gerardo Herrera, Bibiana Martínez, and Francisco Alcantud. "Web Accessibility for Individuals with Cognitive Deficits: A Comparative Study between an Existing Commercial Web and Its Cognitively Accessible Equivalent." *ACM Transactions on Computer–Human Interaction* 14, no. 3 (2007): 1–25.

Sexton, Richard. "The Origin of alt.sex." *Living Internet.* 1995. Accessed 2 January 2009. http://www.livinginternet.com/.

Slatin, John M. "The Art of ALT: Toward a More Accessible Web." *Computers and Composition* 18 (2001): 73–81.

Slatin, John M., and Sharron Rush. *Maximum Accessibility: Making Your Web Site More Usable for Everyone*. Boston: Addison-Wesley, 2003.

Spinuzzi, Clay. "Texts of Our Institutional Lives: Accessibility Scans and Institutional Activity: An Activity Theory Analysis." *College English* 70, no. 2 (2007): 189–201.

Tregaskis, Claire. *Constructions of Disability: Researching the Interface between Disabled and Non-disabled People*. London: Routledge, 2004.

Vanderheiden, Gregg, Loretta Guarino Reid, Ben Caldwell, and Shawn Lawton Henry, eds. "How to Meet WCAG 2.0." *World Wide Web Consortium (W3C): Web Accessibility Initiative*. 1 December 2008. Accessed 3 January 2008. http://www.w3.org/.

4   ENGLISH < A >

JEFF RICE

<A>, *arguably the foundational tag of the Web, provides the hypertext link. On the Web,* <A> *typically provides a one-way link between a fragment of text or an image to another page. Ted Nelson, who coined the term* hypertext, *imagined that links would be multidirectional and multifunctional. He has criticized Web links for their simplicity when compared to Xanadu, his initial hypertext design that has never been completed. In many ways, the one-way link is the genesis of* <A>*'s power. Any site can be linked to any other site without complication. Google's page ranking algorithm capitalizes on this point by comparing the number of links to a given page with other hyperlinked documents. "Deep links," which link directly to specific pages rather than a site's main page, have been the focus of legal cases. In one famous instance, Ticketmaster sued Tickets.com for deep links that bypassed Ticketmaster's advertising. Site owners sometimes use technical means to redirect or refuse these unanticipated links, even though links form the core currency of Slashdot, Meta-filter, Digg, Delicious, and a host of other sites that drive traffic, focus discussion, and energize the memes that cascade daily across the Internet.*

<A> PROVIDES THE BACKBONE of Web writing. <A> is the basis of the hyperlink, a part of <A HREF="">, the HTML code that allows the Web to be an interactive writing space from which nonlinearity, site menus, and navigation are created. <A> gave birth to the Web portal, Yahoo and Netscape's initial 1993–94 hierarchal organizations of Web content. The portal granted access to subjects communally tagged as of interest—Arts, News, Regional, Society and Culture, Education—via a click identified as a blue highlighted <A>. As much as <A> organized information for early Web users when both Netscape and Yahoo joined the Web, today it organizes more complex quests for information online. <A HREF=""> is the foundation of all searches; it directs the ways users move from page to

· 49

page, discover information, or have information presented to them in their browser. <A> has also become the space of various online activities, with search being a dominant one. A search engine such as Google can only deliver its results as hyperlinked pages organized around a search's inputted parameters. A user of a search engine finds the results by clicking on one of the hyperlinks and by following the presented information. Behind the search, Google forms relationships among the linked pages, pushing the most relevant or most popular to the top. Google transforms <A> into a space of information relationships. As John Battelle tells the story of Google, it was built out of the concept that the Web could be made to "reveal not just who was linking to whom, but more critically, the importance of who linked to whom, based on various attributes of the site that was doing the linking" (74). Thus, <A> is as much an organizational tool as it is a linking tool. "The power of the Web is in the links," Albert László Barabási writes. "They allow us to surf, locate, and string together information. These links turn the collection of individual documents into a huge network spun together by mouse clicks" (31).

Given all of these preliminary points, the definition of <A> as "creates a hyperlink"[1] or as an "anchor"[2] is misleading because such a definition focuses on one-way or even two-way connectivity, as opposed to the spatial generation of relationships. More than the basic understanding of forming a connection, <A> is not limited to merely creating a link or establishing a foundational point on a Web page. <A> links the Web's pages by creating a vast network of relationships. In doing so, it generates the non-linear sequentiality Ted Nelson declared as being at the heart of his initial vision of hypertext. Nelson imagined <A> as a textual engagement where all information comes together in multiple ways. Nelson imagined information as relationships formed through links. In *Hypertext 2.0*, George Landow picked up on Nelson's concept of the link and studied <A>'s potential as a rhetorical move that produces informational relationships. Landow writes that the purpose of <A> is

> to make connections between texts and between text and images, the electronic link encourages one to think in terms of connections. To state the obvious: one cannot make connections without having things to connect. Those linkable items must not only have some qualities that make the writer want to connect them, they must also exist in separation, divided. (*Hypertext 2.0* 171)

Indeed, connection has become the focal point of online writing, though in ways that extend connectivity's initial definition from creating "a" link.

In addition, current focus on connection significantly alters Landow's belief that information must be separate before it can be connected. In the online world of Web 2.0, <A> does not begin from the viewpoint of separation and the promise of eventual connectivity. Web 2.0 signifies information as always in a relationship, as always embedded with the attributes of <A>.

Web 2.0, as an emergent concept meant to define the proliferation of social software, is based on the idea of information always being in a state of connection: people connecting to people, ideas connecting to ideas, ideas connecting to people and the connections continuing on to other bodies. Web 2.0 is a term defining the ways users generate content via the link instead of making links connect established content.[3] A chart composed by Tim O'Reilly (cofounder of the term *Web 2.0*) outlines the shift from a supposed Web 1.0 environment to this newer space. Whatever was prevalent in Web 1.0 because of the two-directional hyperlink connection that used separation (i.e., an MP3 file, a Web site, a directory) is now refigured as a multilinking social space (file sharing, blogs, folksonomy tags).[4] The justification for this breakdown, O'Reilly writes, is how Web 2.0 operations view <A> as a collective process or type of collective wisdom:

> Hyperlinking is the foundation of the web. As users add new content, and new sites, it is bound in to the structure of the web by other users discovering the content and linking to it. Much as synapses form in the brain, with associations becoming stronger through repetition or intensity, the web of connections grows organically as an output of the collective activity of all web users.

Despite the hyperbole that often accompanies discussions of Web 2.0, within its myriad developing and developed applications, what began as a focus on <A>'s ability to forge connectivity within the hyperlinked Web page has changed. In that previous definition, <A> is a connector of text or image to other text or image. Within social software, <A> becomes the backbone of larger, database-driven systems where connections occur simultaneously at various levels of experience and various physical, rhetorical, and conceptual spaces. Connections, in other words, socialize. Joining text and image may still occur, but it is not the focal point of the Web activity. Facebook, Flickr, Delicious, Twitter, and popular content management systems such as WordPress and Drupal contain layers of connections dependent on hyperlinks, some of which are visible and some of which are not. The links, as one body and not as separate, divided bodies, make the application run because of how they socialize with one another.

In this context, *social* does not mean "friendly" or "neighborly." It indicates a space of connectivity that situates relationships as a central point. By that definition, the social is not a fixed community; rather, it is the formation of a body of information. It is the formation of relationships.

The social, Bruno Latour claims, "is the name of a type of momentary association which is characterized by the way it gathers together into new shapes" (65). The social is not a permanent body of information; it is a gathering, and gatherings alter the bodies that inform one another. "What lies in between these connections?" Latour asks (221). In other words, what else, besides connecting two or more things, does <A> allow for in a social space? At any moment, the Flickr set may change as new tags are added, the Facebook profile may readjust to new friends or added applications, the comments section of a Metafilter exchange may cause the authors of the initial post or other posters in the comments to shift position. What functions as connectivity is not a direct link in any of these engagements; it is instead the repurposing of <A> as sociality. The social aspect of social software, therefore, makes <A> something other than a solitary linking tool. <A>, in or as the social space, does link, but its linkages do more than connect. They cause shifts. <A> is the basis of extended, continuing, and dislocated information relationships. "Everything is deeply intertwingled," Nelson wrote long before the notion of social software (166). "The *structure of ideas* are not sequential. They tie together every which way. And when we write, we are always trying to tie things together in nonsequential ways" (29). Social software, dependent on PHP and databases in order to function, emphasizes the role <A> plays in forming socialized, intertwingled relationships that remake each other as they intertwingle. In other words, <A> is a social tag. <A> is the physical, technological, and conceptual space where ideas, people, places, and things all become intertwingled and tied to together in a variety of ways.

Ted Nelson might not agree that his plan for a multiconnected online world of information has been realized, to some extent, in social software. Still, much can be said about <A>'s role as the backbone of this emerging space. Without <A>, Facebook cannot connect people and the applications they play with. Without <A>, a Weblog cannot connect its posts internally and externally, nor can a writer even use a Weblog, because every function within the Weblog's space is a hyperlink. The socialness of social software, therefore, intrigues me, for it promises <A> as more than the keyword of a linkable space. It promises a space of continuing connections and relationships. In particular, <A> promises a network of writing where connectivity extends the link. Links remain important, but rather than only connect, they become meshed spaces. If all that matters is connecting, the Web already would be a massive social space of meshed spatial relationships. Barabási argues it is not: "for various technical reasons the links of the Web are direct. In other words, along a given URL we can travel only in one direction" (165).

My interest in exploring <A> as a keyword is in making it travel in all directions at once. As if this chapter is itself a networked, social space, an initial connection I want to make as I consider the role of <A> in writing is the relationship between new media (represented by markup) and the history of writing and education. I do so through an alphabetic traveling, one that moves from the letter "A" to <A>. University writing instruction's origins can be traced to the pivotal moment in 1885 when Harvard University instituted English A, the first freshmen composition course. Designed to fix the sudden problem created when, coinciding with increased admissions, a significant number of students accepted to the university were unable to pass an entrance writing exam, English A marks the moment where universities felt compelled to offer basic writing instruction. English A, however, also marks a new media transitional moment. Those students deemed basic writers by their failure of the exam were graded for not understanding the conventions of print culture: spelling, punctuation, idea organization, and even handwriting. Harvard's assessment shift—from oral recitation to print—necessitated the invention of an alphabetic-oriented writing program, English A. Harvard, in its attempt to socialize education, settled with the concept of "basic" rather than with connectivity and relationships.

The Harvard assessment was based on the notion of the nonrelationship-oriented body, the individual. Individuals, English A's logic dictates, working by themselves on individual ideas or texts, must demonstrate mastery within an individualized space, in this case the essay or exam. The student is thus asked to reproduce the norms and standards of an ideological structure devoted to individuality. At the core of that individuality is the pejorative category of "basic." The English A legacy of individuality as a response to new media (English A is the outcome of the rise of print culture in education) is repeated in contemporary scholarship and commentary. Most recently, Alan Liu's *The Laws of Cool* returns to the Harvard sensibility of the individual by proposing that networked culture has destroyed aesthetic values and replaced them with the most individualized of all experiences, and in some ways, the most basic: cool. "What do the well read," Liu asks, "who once held power in the name of the aesthetic still have to teach the well-informed who now hold power under the cover of cool?" (3). His response is a critical breakdown of everything that networked culture and new media have caused to fail: social values, morals, art appreciation, labor equality, the individual. Liu's vision of new media driven by hyperlinks, among other technologies, is antisocial.

I resist Liu's negative assessment that the world of new media and the Web is the world where information is "designed to resist information" (Liu 179). In his overview, Liu distorts Marshall McLuhan's concept of the global village, the space "when information is brushed against information," by failing to cite the second half of McLuhan's famous dictum, "the results are startling and effective" (*Medium* 76–78). By focusing only on the "brushing" of information and not the outcome of information coming together in a relationship, McLuhan's declaration is reduced to a bemoaning of "the faceless work of acquiring, exchanging, merging, delivering, and otherwise interfacing information" (Liu 185). McLuhan's point was that the socialness of connectivity (posed as juxtaposition, not links) generates responses that we may or may not appreciate, but that are effective for the various kinds of results eventually produced. The protocols of acquiring or merging, therefore, are not the end result of a connection; relationships are the result because of the various effects produced. McLuhan's sense of cool, then, is not an individualized experience, as Liu believes, but rather a relationship-based experience that results in effects (or what he also describes as a participatory process). Instead of the faceless, relationless information economy Liu professes, in this chapter, beginning with McLuhan's initial observation, I argue that <A> does the opposite of destroying information: it builds a world of information within relationships because of how the hyperlink has itself formed a complicated relationship with connectivity. My thinking is an extension of World Wide Web founder Tim Berners-Lee, whose semantic Web is a space where "machines become capable of analyzing all the data on the Web—the content, links, and transactions between people and computers" (157). Although Berners-Lee's focus is on <A>, it is not on the traditional portrayal of the hyperlink's singled-sided connectivity or clickability. As information relates to other information, Berners-Lee writes, "filenames should disappear; they should become merely another form of URI [uniform resource identifier]. Then people should cease to be aware of URIs, seeing only hypertext links. The technology should be transparent, so we interact with it intuitively" (Berners-Lee 159).

Liu does not yet believe in that intuition, metaphoric or actual. At one point in his denigration of new media, Liu mocks the hyperlink.

> "Click" on the "link" on this "page," an ordinary Web page might thus say, unconscious that pages do not click in either the aural or tactile senses and that links in any case would produce a sound more like "clink." (229)

If links do indeed call out to us, as Liu's interpellative critique suggests, might they not ask us to "click here," but instead to form a relationship? That is, might the link not stop with the superficial gesture of a click but instead focus on what happens next, what relationship is formed? If anything, educational practices, which Liu fears are being undermined by the Web, have not stressed the relationship aspect of information organization as much as they have stressed the metaphoric "click here" aspect. In the "click here" gesture, a student answers correctly, sees content as situated in a predefined place, and produces work that resonates with a thudlike click. Writing, in the Liu imagination, is a click. In the meta "click here" moment, this thinking suggests, Harvard's English A is founded. It, too, asked writers to "click here" by performing basic tasks in writing instruction that often resulted in a type of clink. To begin my discussion of <A> as a social link, I want to further this idea by contextualizing <A>, the subject of Liu's overall scorn, with the origins of a specific "click here" thud, Harvard's English A.

English A

When the university writing paradigm was instituted as a print-oriented concept in the late nineteenth century, the alphabetic and hierarchal structuring that print generates situated first-year writing, or writing instruction in general, as a service, an occupant of the lowest level of academic work. In their 1892 "Report of the Committee on Composition and Rhetoric," Charles Francis Adams, Edwin Lawrence Godkin, and Josiah Quincy described English A's purpose as follows:

> "English A,"—the course prescribed for the Freshman class—is designed to give (1) elementary instruction in the theory and practice of English Composition, and (2) an introduction of the study of English Literature. The theory is taught throughout the year by lectures; the practice is obtained in short

weekly themes, written in the class-room and criticized by the instructors. (qtd. in Brereton 75)

The very nature of its title, A, indicated the elementary status these educators attributed to writing instruction. Alphabetic literacy, as McLuhan recognized, instills a sense of hierarchy in our ability to organize information and social structure. Alphabetic literacy, represented in the A of English A, situated writing in terms of categorization, definition, and separation. This gesture positioned writing and rhetoric, within the educational apparatus, as a basic skill disconnected from the scholarly demands other academic pursuits require or that might be performed in a given space (a lecture, for instance, followed by practice). As John Brereton tells the story of Harvard's English A, "If the course came in the freshmen year, it would help the student when he most needed help, and it would not break into his other studies when he had taken them up" (30). English A's origins argued for separation, therefore, at the level of content (this course would be distinct from other areas of study by not breaking into them) and rhetorical value (what is accomplished in the course belongs only in the course). English A also argued for separation of the student from other parts of the student body (the course would not be integrated with major studies of learning). By following these rules, one could assume even in the late 1800s, English A, as a body of information, would never change. Without institutional interaction, it couldn't.

The examination for acceptance into the university that led to this move came with the 1873 Harvard admission requirement for English proficiency:

Each candidate will be required to write a short English Composition, correct in spelling, punctuation, grammar, and expression, the subject to be taken from such works of standard authors as shall be announced from time to time. The subject of 1874 will be taken from one of the following works: Shakespeare's Tempest, Julius Caesar, and Merchant of Venice; Goldsmith's Vicar of Wakefield, Scott's Ivanhoe, and Lay of the Mast minstrel. (qtd. in Brereton 34)

This model, with its emphasis on standardization, rote performance, and subjection to the "great" individual author also indicates the origins of nonsocial composing. As important as it seemed to English A's authors to not allow writing to interfere with other studies or students, it also seemed equally important to equate writing with the notion of one individual working independently; similarly, the single author the student studies works independently (and is read as a single body of information). Two levels of individuation were emphasized.

Granted, it is easy to offer this critique over a hundred years later. The point of the critique, however, is not to find fault with Harvard but to situate the hyperlink's legacy with that late nineteenth-century moment via an initial alphabetic connection, A. English A's recognition that late nineteenth-century writing instruction had not adapted to changes in communication innovation (print) provides an analogy as well as a social connection for contemporary education practices. What the rise of individuation did for education is tremendously important. Harvard's exam evaluators found serious problems in print-related, surface-level issues that would not have existed in oral delivery of school subject matter: standardization in punctuation, article placement, and even handwriting. Failure to master any of these or other skills gave cause for failure. In his 1963 *Themes, Theories, and Therapy*, Albert Kitzhaber describes the problems associated with this exam as the first moment the university sees writing instruction in terms of basic skills—that is, writing teachers perform a necessary service for the university by teaching students punctuation, grammar, and style. Kitzhaber noticed the pedagogical problems inherent in such attention. A year later, McLuhan made the pedagogical individuation that Harvard created the focal point of his interests in media:

> At school, however, [students encounter] a situation organized by means of classified information. The subjects are unrelated. They are visually conceived in terms of a blueprint. The student can find no possible means of involvement for himself, nor can he discover how the educational scene relates to the "mythic" world of electronically processed data and experience he takes for granted. (*Understanding Media* viii–ix)

"Involvement," McLuhan claimed, concerned the social aspect of media connectivity. As much as my initial outline might appear to trivialize Harvard's historical instruction, and as much as my argument here works off McLuhan's critique, individuation's place in contemporary curricula has been one of value even if current new media developments complicate the practice. If print (represented by A) stressed the individual, as McLuhan also claims, individuals interpellated by print culture needed to learn how individuation plays a role in communicative practices. The contemporary dominance of <A> (as opposed to that of A) is a call for an update of this awareness. Thus, one might ask whether the emergence of the social, via <A>, poses similar consequences for contemporary pedagogy?

Social Networks/Social <A>

There should be little surprise that the English A curriculum, as an educational protocol, has had long-term ideological consequences. Valuing writing as a private, individual experience ensures emphasis in correctness, a necessary skill within a variety of communicative practices, but also deemphasizes the need for relationships. In the English A model, writers and writing content are considered to be independent bodies. "The way we talk about a subject becomes part of the subject," Gerald Graff writes in his argument for a revised college curriculum. "Students must not only read texts, but find things to *say* about them, and no text tells you what to say about it" (9). As English A once taught, Graff asks pedagogy to admit that the student (one body) and the text (another body) are independent of one another and not connected. No text tells you what to say. If it did, the writer would have to be in a relationship with that text. Such a declaration is driven by print protocol's "specialist purposes often serving higher cultural purpose" (i.e., university education) (Goody 77). Graff's plan of action is more alphabetic than networked. What he names "street smarts" or "hidden intellectualism" is nothing more than an emphasis on individual identity. Graff admits to this logic of alphabetization in an anecdote from his own freshmen writing teaching: "As many [of Graff's students] saw it, 'English' was one more hurdle among others, with no apparent relation to anything else they studied. I was one more incomprehensible prof to be humored and circumvented" (67). The tone is self-mocking, but its message repeats English A's initial documentation. In other words, the ideology of keeping informational bodies separate in education extends from the 1870s to the twenty-first century.

In their exhaustive analysis of networks, Alexander Galloway and Eugene Thacker argue that the ideology of Web protocols (as opposed to those associated with print) affect political, economic, and even biological work by dislocating the individualized and specialist experience for a networked one where individuality (often understood by the label *sovereignty*) is destabilized. For Galloway and Thacker, the problem is not that bodies are kept separate, but that specific network connectivity decenters certain power structures at the expense of others. Web protocols, they insist, are layered activities that form relationships with one another as this decentralization occurs.

> Technical protocols are organized into layers (applications, transport, Internet, physical); they formalize the way a network operates. This also allows

us to understand networks such as the Internet as being more than merely technical. (42)

As much as Galloway and Thacker are concerned with the implications of such formalizations, particularly ones that establish controlling mechanisms, educational practices are missing from their discussion. Although linking has allowed "peer-to-peer file-sharing networks, wireless community networks, terrorist networks, contagion networks of biowarfare agents, political swarming and mass demonstration, economic and finance networks, online role-playing games, personal area networks, mobile phones, 'genreation Txt,'" and others to proliferate (as they write in their opening pages), education, too, I note, deserves such attention. The challenge of protocols, Galloway and Thacker write, "bears as much on cultural theory and the humanities as it does on computer science, molecular biology, and political theory" (99). One might assume, then, that despite its absence in Galloway and Thacker's list, education is also being decentralized as other networked forces gain power. In their reading, <A> should be a culprit, not a Ted Nelson–inspired pedagogical tool. Without disagreeing with all of Galloway and Thacker's analysis of networks, I want to argue that as a social process, <A> is more likely to function as a protocol that does not bear down on these various activities or areas, but instead that puts each into relationship with the other in an ever-assembling process. This process, Latour writes, is an indicator of "the number of associations needed to carry out even the smallest gesture" (242). Associations may lead to control as they assemble, but they also may lead to new bodies of information in general. If anything, a network, as well as its various interacting and breaking-apart nodes, could be the very nodes Galloway and Thacker's three identified areas (plus others) consist of, even if the writers have not yet visualized this social aspect of <A>. A network could be the basis of an educational practice.

If we are going to call the potential network they do sketch by any name, we might as well call it education. At least, what I call here education is one network driven by the new media logic of <A>. Education in this sense includes a number of associations that assemble various nodes, though they do not do so by order or strict adherence. The nodes may be exchanged, fall out, increase in size, or change in other ways. This network does not speak to one area of study or methodology, but it does recognize writing as the driving feature of <A>'s characteristics. George Landow hints at such a model when he proposes hypertext as the basis for a new media educational system. Landow recommends the link as writing that joins educational practices. "Since links cross borders and reconfigure our senses of relationships, why not use them to reconfigure the relations of museum and university?" Landow asks ("Paradigm" 40). He then shifts attention to the university alone:

> I emphasize that this examination of the relationship between hypertext, an important form of digital information technology, and university education focuses chiefly upon hypertext as a paradigm, as a thought-form, rather than on the tails of hardware and software. ("Paradigm" 40–41)

As a thought form, the metaphoric "click here" gesture I borrow from Liu's critique of new media might be accused of non-<A> logics, for the "click here" Liu highlights is a one-way process. As a thought form, <A>, on the other hand, might be the basis of some kind of new media education structure whose protocol is in the relationship. As Landow suggests, <A> utilizes "the way our information technologies permeate the way we think" ("Paradigm" 44). McLuhan argued that without new media thought forms, "we think" from a nineteenth-century position that "characterizes the educational establishment where information is scarce but ordered and structured by fragmented, classified patterns, subjects and schedules" (*Medium* 18). Landow's vision makes information hypertextual, and, we might imagine, relational.

Jay David Bolter argues similarly when he calls on educators to treat hypertext as a pedagogical application because it can reshape various educational practices. Bolter asks for a revitalization of writing that functions hypertextually, though not necessarily on actual computer screens. Bolter situates <A> as the link for generating a social scholarship:

> If new media are becoming accepted in pedagogy, the question remains whether and when humanists will extend their notion of critical research beyond print to include new media forms. Will they be willing to redefine scholarship to include the multilinear structures of hypertext or (what may be even more radical) the multiplicity of representational modes afforded by digital media? (29)

Bolter, however, remains skeptical of <A> accomplishing much because of specific ideological structures and protocols already in place, some of which we can trace to English A. "The theory community seems unwilling to extend its experimentation to electronic forms such as the linked hypertext or hypermedia document," Bolter writes, as if to say that theory (identified as poststructuralism) still fixates on individuation as opposed to a more social, electronic form of writing (28). Still, some academics, he notes, "understand that writing in the new digital environment can be a hybrid form of communication in which words instantiate and inform images as well as the reverse" (20). In this mix of unwillingness (ideas must be expressed in one specific way) and acceptance (we live in the age of new media where at least one kind of relationship, text and image, exists), <A> is not yet utilized to show either the programmatic or scholarly model desired. That model is hinted at in Darren Tofts, Ray Kinnane, and Andrew Haig's "I Owe the Discovery of This Image to the Convergence of a Student and a Photocopier," in which the authors associate <A> as a pedagogy that teaches the basis of Web logics. In a collage of interlinked quotes, insights, and images, they argue for a reproduction of their own attempt to write hypertextually:

> Pedagogy, like hypertext, becomes a non-sequential process, a network of intersecting paths, some established (prescribed reading), some implied (suggested reading), others improvised (chance associations, initiative, idiosyncratic links). (255)

Like Landow and Bolter, Tofts, Kinnane, and Haig explore hypertext as educational. "Interactive technologies," they note, "promise a high degree of intervention and involvement in knowledge production" (25). <A> as hypertextual, the authors claim, is an attractive tool for enacting such a promise. "The individual is dominant until the close of the 19th century," they insist, echoing McLuhan (256). "The presentation of electronic media is one of constant collage" (258). <A> evokes collagist writings, "interesting networks of themes that open up a conceptual map" (258).

Indeed, collage often is the focus of this type of educational/digital education or writing that asks students to merge texts and ideas. <A>, as a social space, though, is not a collage. As a network, <A> builds relationships while also becoming relationships. To enact a pedagogy of <A>, I have to imagine a social software logic as opposed to connection or merging among things. That logic is not bound to a specific platform, such as the Web, but instead informs institutional practices through the generation of large-scale spaces. Continuing my initial relationship between A and <A>, therefore, I call this pedagogy not a collage or hypertext course, but instead, I call it English <A>.

English <A>

English <A> is an extension of the three juxtapositions I previously introduced regarding education and hypertext: Landow's thought form, Bolter's
research, and the pedagogy of Tofts, Kinnane, and Haig. These three ideas,
however, have not yet been socialized as a networked space. As is, they
each propose a hypertextual educational practice dependent on the nonlinear, linking aspect of <A>. One might link texts and images as a new
research practice or pedagogy, but the process of linking a text to a text or
a course in one university to one in another university limits <A>'s social
aspects. As a hypothetical space, English <A> borrows this juxtaposition
I produce while also constructing <A> into a network. The networks are
not spaces where one may study multiple subjects, interlink multiple subjects, or study in nonlinear fashion, because to do that work would rein-
force the individuality of English A that most hypertext theorists want to
move away from. To do that kind of work would make the user (or writer
or course or subject) the individual at the center of all activity, not a part
of the network. Instead, the network I imagine <A> creating is a flexible,
shifting, never stable entity (or even entities). <A> causes these entities
to form and dislocate, becoming larger and smaller bodies as they do so.
Indeed, to follow Latour's concept of the network, the moment one pedagogical space engages with another (which, in essence, would be every
moment), both spaces link and then change on the basis of what and how
they assemble. The terms they use socialize, the ideas they use socialize,
the people they use socialize, and an assemblage occurs. To do an assemblage for education, I engage with the ideas presented here as well as the
logics of several social software programs without calling for actual usage
of these programs. And as I engage with each, I make a larger pedagogical
network. My software choices, overall, are among the most basic applications available, yet each can contribute new meanings for <A>. Each
offers a possibility for <A> that English A could not provide. Each provides a possibility for belonging within English <A>. Each, we might say,
is a tag within the network.

Online, networks that function via hyperlinks without connectivity
being their sole purpose incorporate tags so that users can extend the
given space's organizational possibilities. The tag is a marker, a name or
image that identifies a word, image, site, story, or related content. Tags may
be communal or eccentric. Tags are typically hypertext links. Clay Shirky
offers the following definition of tags:

Tags are simply labels for URLs, selected to help the user in later retrieval of those URLs. Tags have the additional effect of grouping related URLs together. There is no fixed set of categories or officially approved choices. You can use words, acronyms, numbers, whatever makes sense to you, without regard for anyone else's needs, interests, or requirements.

Labeling has been an important gesture for pedagogy and research: an individual uses new media, an individual is in the humanities, an individual produces scholarship. The same kind of labeling can be performed with content produced in each area, where each area is completed, and who or what responds to each area. In this way, key terms are circulated, often independently of one another and as alternatives to the English A pejorative "basic." Individual bodies dominate. Nothing is intertwingled.

Three social software sites dependent on the tag as <A> extend labels as a kind of intertwingling so that writing occurs within and as networks: Delicious, RSS, and Flickr. Delicious, a social bookmarking system, engages <A> at the level of information sharing. Whereas the traditional bookmarking tool found in all browsers allowed Web users to catalog sites for future viewing, Delicious extends the catalog to a public venue via an accessible and shared URL. Users who want to catalog a site also indicate how the site will be tagged; that is, users choose what kind of indicator or name will be attributed to the collected site. Instead of filing the site into a metaphoric folder, the site is given a marker that, in turn, is itself a hyperlink connected to all other users who have either chosen the same marker or bookmarked the same site. In this case, <A> is an archival tool as well as an information-sharing tool. RSS (Really Simple Syndication), too, performs a cataloging function. Instead of bookmarking, however, it provides links to Weblog and content management systems that provide feeds (i.e., updates) to their content. RSS feeds deliver content into the browser and thus work via the concept of relevance. Timeliness aggregates information in one space. That information is often tagged for user relevance. Flickr, on the other hand, juxtaposes visuality and tagging by allowing users to name their uploaded images by any word they desire. Flickr, like Delicious, shifts the collection and storage of information from private to public spaces. A part of the Yahoo network of online holdings, Flickr establishes large-scale and small-scale networks of users through the general search feature and the contact feature. Contacts are users who have a relationship with each other's work. Like the other two applications, Flickr is an organizational tool; images may be grouped via <A> into sets, collections, and batches. Each organizational scheme affects internal and external navigation; each is put into a relationship with the contacts.

This information defines what each application does, but more importantly, it provides some context for how English <A> might utilize social software logics for a potential writing program. I take from each an important aspect of socialized information motivated by <A>'s relationship with tagging: publicly archived, interlinking, and connected research; the relevance, aggregation, and push of a feed; and visually tagged sets within user groups (where a user group's definition is open-ended depending on how users form relationships in a given space). As a larger network, these three items use <A> to produce relational spaces of information. In English <A>, the individual who would previously work by herself on the work of an author or text in a single space in a single institution at a single moment (as English A's exam sets forth) instead works with <A> as a tag so that information is publicly named and engaged with, is pushed toward other users and spaces by tagged content, and is visual. Tagged subject matter, methodology, space, other writers, and even one's self are put into the same composing processes (along with other areas to be tagged) in order to allow for new media writing that is relational. What I tag, what I push to another space, and what I visualize enter into visual and textual relationships with other material, but also with me, with another individual, with a larger group, with an application, with a practice, and so on. In this context, relational means that <A> is the basis of a large network of people, things, ideas, and places that find connectivity rather than are made to connect. Relational means one does not write about a subject or author and then link to another subject or author. It means that one tags spaces via <A> so that they can be socialized with other spaces.

And in that final definition, we might imagine the program English A cannot generate. A is a marker of a single letter; it produces single writers (basic or otherwise) in single spaces (classrooms) on single texts (paper, exam). English <A>, on the other hand, motivated by tags, associates and assembles across numerous spaces in numerous texts. In English A, the student resembles the letter A, an independent body that could be placed with other bodies to create meaning, but that mostly subsists on its own. As a tag, however, <A> allows multiple moves across multiple spaces so that multiple bodies are created. The student, the text, the word, the image, the space, and so on are tagged in relationships. From the set of ideas to be used in a given rhetorical situation (learning from Delicious) to getting those ideas labeled in a variety of ways and delivered to an audience or audiences for specific moments (RSS) to visualizing those ideas as identities, groups, things, places, or some other body or set (Flickr), <A> is no longer bound to unidirectional or bidirectional movement. Movement, we discover, is everywhere at once. What I propose as a keyword, therefore, is a thought form regarding research, writing, and pedagogy. As a thought form, I do not detail a specific moment or program but rather the possibility of many such moments, programs, and movements. Although these three applications can be used for the purpose I am imagining, the thought form I initiate is meant to go beyond the software itself so that the logic of the software motivates further work. Unlike the early Web portals that used <A> to facilitate unidirectional navigation of sites of interests, <A> as social network eliminates the entry keyword (Arts, Education) by making the process always ongoing and always pushed toward new moments. Generalizing from this shift, it no longer makes sense to tag writing or writing instruction as the unidirectional tag that sends the student through other courses or a larger institution. The socialized experience makes writing always relational.

Notes

1. See the once-popular Webmonkey HTML guide: http://www.webmonkey.com/reference/html_cheatsheet/.

2. See the World Wide Web Consortium definition: http://www.w3.org/TR/html401/struct/links.html.

3. I am aware of the essential roles Ajax and other backbones such as Ruby on Rails play, but the heart of Web 2.0, I will argue here, derives from how each has changed the content and performance of the link or <A>.

4. See http://www.oreilly.com/pub/a/oreilly/tim/news/2005/09/30/what-is-web-20.html for the full chart and explanation.

Works Cited

Barabási, Albert László. *Linked: The New Science of Networks.* Cambridge, Mass.: Perseus Books, 2002.

Battelle, John. *The Search: How Google and Its Rivals Rewrote the Rules of Business and Transformed Our Culture.* New York: Portfolio, 2005.

Berners-Lee, Tim. *Weaving the Web: The Original Design and Ultimate Destiny of the World Wide Web.* New York: Harper-Business, 2000.

Bolter, Jay David. "Theory and Practice in New Media Studies." In *Digital Media Revisited: Theoretical and Conceptual Innovations in Digital Domains,* edited by Gunnar Liestøl, Andrew Morrison, and Terje Rasmussen, 15–33. Cambridge, Mass.: MIT Press, 2003.

Brereton, John. *The Origins of Composition Studies in the American College, 1875–1925: A Documentary History*. Pittsburgh: University of Pittsburgh Press, 1995.

Galloway, Alexander, and Eugene Thacker. *The Exploit: A Theory of Networks*. Minneapolis: University of Minnesota Press, 2007.

Goody, Jack. *The Domestication of the Savage Mind*. Cambridge: Cambridge University Press, 1995.

Graff, Gerald. *Clueless in Academe: How Schooling Obscures the Life of the Mind*. New Haven, Conn.: Yale University Press, 2003.

Kitzhaber, Albert. *Themes, Theories, and Therapy: The Teaching of Writing in College*. New York: McGraw-Hill, 1963.

Landow, George. *Hypertext 2.0*. Baltimore: Johns Hopkins University Press, 1997.

———. "The Paradigm Is More Important Than the Practice: Educational Innovation and Hypertext Theory." In *Digital Media Revisited*, edited by Gunnar Liestøl, Andrew Morrison, and Terje Rasmussen, 35–64. Cambridge, Mass.: MIT Press, 2003.

Latour, Bruno. *Reassembling the Social: An Introduction to Actor-Network-Theory*. New York: Oxford University Press, 2005.

Liu, Alan. *The Laws of Cool: Knowledge Work and the Culture of Information*. Chicago: University of Chicago Press, 2004.

McLuhan, Marshall. *The Medium Is the Message*. 1967. Reprint, San Francisco: Hardwired, 1996.

———. *Understanding Media*. New York: Signet, 1964.

Nelson, Ted. *Computer Lib/Dream Machines*. Redmond, Wash.: Tempus Books of Microsoft Press, 1987.

O'Reilly, Tim. "What Is Web 2.0: Design Patterns and Business Models for the Next Generation of Software." 30 September 2005. Accessed 15 June 2007. http://www.oreilly.com/.

Shirky, Clay. "Ontology is Overrated—Categories, Links, Tags." Accessed 10 October 2007. http://shirky.com/.

Tofts, Darren, Ray Kinnane, and Andrew Haig. "I Owe the Discovery of This Image to the Convergence of a Student and a Photocopier." *Southern Review* 27 (September 1994): 252–60.

5 A STYLE GUIDE TO THE SECRETS OF <style>

Brendan Riley

Style appears in multiple forms: style *is both an element placed in the head of documents and an attribute that can be added to many other elements such as* <p>, <img>, *or* <table>. *What* style *adds, in either application, are the graphical and typographic directives of Cascading Style Sheets. But styles live in a liminal world: they are to some extent separated from the content and structure of an HTML document. Although early Web pages might have conveyed this information via repeated tags such as* <b>, <font>, *and* <table>, *browsers enable and standards-compliant development practices encourage imagining* <style> *as optional. Indeed, the content of* <style> *can be ignored, replaced, or remixed by end users. An early* <style> *trick was to give readers the option to change a Web site's color or layout (clicking on the link calls a new style sheet). RSS newsreaders, on the other hand, often present the content of Weblogs (their text and images) without the colors, backgrounds, and branding supplied by* <style>.

> If you want to know why we're on top of the pie? Cuz we got style.
>
> **GRANDMASTER FLASH,** "STYLE (PETER GUNN THEME)"

THE <style> TAG IS UNIQUE IN HTML (Hypertext Markup Language). Although many tags perform singular functions, no other tag has such a wide range of effects or power as the <style> tag. Indeed, <style> often contains more information than the rest of the HTML file. Web authors may even place its contents into an external document, a style sheet. Although this is certainly convenient, this separability embodies a key duality at the heart of the evolving digital sphere.

In dividing style from the remaining content, the World Wide Web Consortium (W3C) reinforces a division long held by designers and writers. When William Strunk, coauthor of *The Elements of Style,* suggests that writers must first learn to follow rules before they write with their own voice, he is suggesting that style matters less than content. One can craft one's own style, but only if it doesn't conflict with the established norms. HTML's hard-coded separation between style and content similarly positions style beneath content; when browsers can supersede style or turn it on and off like an amusing feature, style certainly seems less important.

In an age where most of our style choices are made by default, the Web's stylistic smorgasbord leaves those focused on alphabetic content reeling—and those trained in formal visual design retching. Rather than carve off style as something to be left to the professionals or ignored completely, there might be another approach that considers how writers can style their texts. This chapter, a style guide of sorts, seeks just such an approach.

Cascading Style Sheets

Ironically, the medium most widely recognized for changing our communicative habits amplifies the split between content and style. The evolution of the `<style>` tag mimics quite nicely style's shifting rhetorical value in the age of electracy.

HTML was originally created as a simple, fast way to post shared information. Its roots in science and technology downplayed visual design, focusing instead on cross-platform compatibility, uniformity, and efficiency. But as Web browsers evolved, they added their own tags to make visual design more consistent. However, the accumulated mess of tags made the formerly readable HTML code inscrutable: "HTML tags were originally designed to define the content of a document. . . . As the two major browsers [added new tags], it became more and more difficult to create web sites where the content . . . was clearly separated from the . . . layout" (W3Schools). The W3C responded by adding styles, and their more advanced counterparts, Cascading Style Sheets (CSS), to hold the form/content distinction firm. Although many characteristics could now be added to a page element (such as name, hyperlink, or alternative text), style now stood in its own category, governed by different rules and carrying a different weight. Creating a separate rule set gave designers a freedom they had been missing, but it also reinforced the original focus of HTML—the information. Protocols such as HTML, XHTML, and CSS, dubbed *standards,* encourage designers to build their sites so that the reader, not the designer, can control the site's look and feel.

For these standards to work, designers and browser companies must follow the W3C's rules. Compliance advocates shape their arguments around access and uniformity, maintaining the hierarchy of content over form. As the Web Standards Project (WaSP) authors explain:

> Though our message initially met with resistance (particularly from browser companies' marketing and P.R. departments), eventually we prevailed—in part because engineers at many browser companies agreed with us and saw WaSP as an ally in their internal battles with management.

The authors' smug parenthetical conjures hard-working Dilberts frustrated by dim-witted bosses: if only marketing and public relations people understood the importance of structural uniformity and efficiency, we wouldn't have all this trouble! Another telling passage explains the gaps between engineering and marketing:

> Many practitioners take pride in delivering sites that look and work exactly the same in compliant and non-compliant desktop browsers alike, at the cost of accessibility, long-term viability, forward compatibility, and lack of alternative device support. Others develop proprietary code that works only in a handful of popular browsers.

WaSP members suggest these costs are heedlessly paid. They ignore the possibility that designers choose to pay such costs in order to maintain control over their product's design. For WaSP standardistas, rigid protocols ensure portability and uniform access to content; design, now flexible and perhaps even fun, is just a nice add-on.

But style doesn't separate so easily. For instance, many designers use in-document <style> tags to lay out their information. In doing so, the designers embed form right back into content. Context-specific tags, such as <span> and <div>, exacerbate the issue, returning the style code to the sentence level. Even where designers enforce rigid separation between form and content instructions in their code, the traces of style sheets remain sprinkled throughout in class and id attributes.

Legacy browsers pose a more significant problem. From the perspective of split form and content, legacy browsers should not be an issue. Pages meeting standards function because browsers unable to handle newer instructions use default settings, rendering essential content without the more elaborate form. However, the legacy browser's distorted design can damage the site's message as much as jumbled letters would. This second perspective results in a more elaborate approach to legacy browsers. The designer must choose between using competing tags, avoiding any non-standard tags, and producing alternate pages for each browser. The more visual control the author seeks, the more carefully she will need to test her site with many browsers.

The history of testing for different browsers and serving alternate style sheets highlights the rhetorical potential of style sheets. In February 2003, the creators of the Opera browser accused MSN.com of trying to undermine it by sending faulty style sheets to Opera users. If the Opera user instructed the browser to identify itself as Internet Explorer, the style sheets served were free of errors and worked just fine (Lie). It seems likely that MSN.com's scheme springs from the long history of testing for, and sending out, browser-specific style sheets.

This incident highlights the integral relationship between content and form in the experience of Web users. Although code is used to enforce the content/form split, the W3C ignores the nature of style sheets as they appear in the wild. Users with mainstream, recent-model browsers likely do not know about the codified separation; they experience a unity between content and form. The irony of this smooth integration is that the W3C's decision depends on a philosophy that cleaves them. Tim Berners-Lee explains,

> The primary principle behind device independence, and accessibility, is the separation of *form* from *content*. When the significance of a document is stored separately from the way it should be displayed, device independence and accessibility become much easier to maintain. (168)

The W3C externalized design to remind designers that pages are information; content is king.

This philosophy remediates thought patterns developed under print literacy. As Marshall McLuhan explains,

> Perhaps the most significant of the gifts of typography to man is that of detachment and noninvolvement—the power to act without reacting. Science since the Renaissance has exalted this gift. . . . It was precisely the power to separate thought and feeling, to be able to act without reacting, that split literate man out of the tribal world. (173)

For McLuhan, print literacy fosters distinctions between thought and feeling, idea and emotion, content and form. In reifying these distinctions, the W3C upholds the philosophies of enlightened rationalism and print literacy. Like McLuhan, many scholars have noticed that print texts are moving away from linear, language-centered forms toward spatial, visual forms. Gunther Kress describes this process: "Now writing is not the vehicle for conveying all the information which is judged to be relevant. Here language is implicitly seen as a medium which is only in part able to express and represent what needs to be represented" ("English" 74). The rising importance of the visual was a slow process in print. Hypertext, however, was a different story: despite the W3C's best efforts, the Web quickly adopted multimodal forms.

Style Guide

Style guides shape writers in two ways. The most direct method, prescription, involves an intricate set of rules the writer must follow. Web standards such as CSS and stylebooks such as the *Associated Press Manual* rely on this model. Proscriptive guides such as *Style: Ten Lessons in Clarity and Grace* and writing handbooks such as *The Little English Handbook* function negatively, rejecting "bad" writing and explaining how to avoid it. Both models craft boundaries for writers, and both depend on the split between form and content.

Prescriptive style guides provide audience-targeted rule sets that explain how to use specific elements in specific ways—how to frame citations, whether to underline or italicize, and whether to use the Oxford comma, as I just did. Such guides mark with rigid limits the discourse community's boundaries; they provide authority via consistency. Institutional stylebooks are the most common form of prescriptive guide: the *Associated Press Stylebook and Libel Manual*, for example, includes the proper way to write brand names such as *Coca-Cola*, use abbreviations such as *SAM* ("acceptable on second reference for *surface-to-air-missile*"; French 184), and how to cover rodeos. Many organizations establish house styles by endorsing a specific stylebook and establishing additional rules regarding mission-specific data. Such guides foster organizationwide consistency in language and tone, allowing writers to focus on issues of content. At the same time, stylebooks whittle down the choices writers can make about their writing. Their authoritative tone brooks no variance. CSS brings prescriptive style control to Web design; external style sheets enforce a technological house style directly across all the documents using that style sheet. Unlike systems relying on traditional stylebooks, however, CSS allows the designers to change house style at any time, instantly, for all the documents in the site.

Prescriptive style guides define rules to follow. Alternately, proscriptive guides function negatively, aiming at "bad" writing and embracing practices that avoid the condemned styles. For instance, Joseph M. Williams's *Style: Ten Lessons in Clarity and Grace* focuses on general practices—not specific rules—to avoid unclear writing. From the outset, it contemplates writers who craft unwieldy, difficult sentences:

> We might say that such writers have a problem with their style, but they don't, because they have no style, if by style we mean how we choose to arrange our words to best effect: They do not *choose* how to write, any more than they choose to put *the* before *dog* rather than after. But choice is at the heart of clear writing. (4)

For Williams, style immediately equates to clarity; writers who do not choose clarity have no style. As the reader learns to edit for clarity, he must adopt Williams's own house style: transparency. Significantly, Williams focuses on applying style at the end of the process (not unlike Web authors applying an external style sheet). In *The Craft of Research,* the authors urge writers to "revise for style, preferably toward the end of the process" (Booth, Colomb, and Williams 215). Revising for style excises unwanted text, bringing polish to the finished content. On the Web, proscriptive style guides often function as "design don't" lists. For example, Web designer Stefan Mischook offers a thirty-element list of Web design "Do's and Don'ts," more than half of which are don'ts. Eight of the first nine, in fact, are proscriptions.

Prescriptive and proscriptive methods meet in the most pervasive of style guides, the English composition handbook. These texts serve as information centers for writing students and are often proffered as essential desk furniture for the productive learner. Many colleges require (prescribe) that every composition student must buy a handbook and work with instructors to learn how to use it. On the one hand, these texts prescribe grammar and citation practices for students, setting the boundaries for appropriate academic style. For example, Elaine P. Maimon and Janice H. Peritz include extensive sections on grammar and sentence structure in *A Writer's Resource.* They also provide detailed guides to MLA- and APA-style citation. On the other hand, composition handbooks also generally include proscriptive sections, discussing the dangers of passive voice or jargon-heavy text. Andrea Lunsford's *The Everyday Writer,* for instance, centers around the twenty most common errors college writers make. Similarly, *The Little English Handbook* lists three disallowed choices and then explains:

> *Choice* is the key word in connection with style. . . . Grammar will determine
> whether a particular stylistic choice is *correct*—that is, whether a particular
> locution complies with the conventions of the language in which we are writ-
> ing. Rhetoric will determine whether a particular stylistic choice is *effective*—
> that is, whether a particular locution conveys the intended meaning with the
> clarity, economy, emphasis, and tone appropriate. (Corbett and Finkle 55–56)

The authors continue with advice about careful word choice, awkward
sentences, and passive voice. Although these proscriptions stress the freedom of choosing one's words, their stress on appropriateness reminds the
reader that choices carry consequences; absent are any discussions of
individual voice or character. The twin constraints of proper grammar
and effective rhetoric encapsulate the handbook's notion of style.

Whether prescriptive or proscriptive, style guides imagine style as an
ornament, a configurable variable presenting content in its best light,
something for individual expression or, that must be kept in check by
clarity and grace. Fortunately for us, with new media comes new style.

Efficiency

Style guides for the Web uphold clarity as essential to good writing. The
Web's adherence to this style stems partly from its roots in hacker culture.
When early programmers first tinkered with computers, technology's limits required that they be as efficient as possible. Programmers sought the
"right" way to code, praising those who shaved lines from programs (Levy).
This Fordist efficiency carried over to the Web, where unadorned presentation reigned supreme. Hacker standard-bearers such as Eric Raymond
and organizations such as GNU.org maintain these priorities: Raymond's
Web site, for example, embraces with nostalgia the unadorned text of the
early Web.

But design needs not be so distinct from content. Nearly a decade before the World Wide Web went mainstream, Donald Norman suggested a
more even-handed approach to the aesthetic/usability split. He writes in
The Design of Everyday Things:

> If everyday design were ruled by aesthetics, life might be more pleasing to the
> eye but less comfortable; if ruled by usability, it might be more comfortable
> but uglier. If cost or ease of manufacture dominated, products might not
> be as attractive, functional, or durable. Clearly, each consideration has its
> place. Trouble occurs when one dominates the others. (151)

Although Norman's book leans heavily toward usability as the keystone of good design, this nod to the balanced need of each element gives his text a sense of balance itself. By contrast, recent focus on usability and calls for standards from groups such as WaSP drift away from aesthetics. Jakob Nielsen's *Designing Web Usability* emblematizes this approach. He writes:

> There are essentially two basic approaches to design: the artistic ideal of expressing yourself and the engineering ideal of solving a problem for a customer. This book is firmly on the side of engineering. While I acknowledge that there is a need for art, fun, and a general good time on the Web, I believe the main goal of most web projects should be to make it easy for customers to perform useful tasks. (11)

Tellingly, Nielsen follows this passage with a list of errors designers commonly make (not unlike Lunsford's twenty errors). He suggests that design errors—poor information structure, design outweighing usability, traditional writing style—stem from a failure to switch from a print to a Web perspective. But Nielsen's Web perspective still presumes a split between form and content; like Williams's *Style,* his approach focuses on clarity and grace. Nielsen relegates style, like the absent personal voice, to ornamentation in service of content.

What if, instead of seeing these two elements as separate parts, like sides of a coin, we imagined them as part of a sliding scale? In *The Economics of Attention,* Richard Lanham reconsiders efficiency's primacy, suggesting that digital technologies make style more clearly valuable. One argument he makes supposes a spectrum of attention based on one's perception: this "perceiver spectrum" shifts from looking through a text to looking at a text. Texts that encourage the former are those that use style as a means to achieve clarity and grace; texts at the other extreme use ostentatious style for their own sake. For Lanham, digital writing opens new flexibility for writers on that spectrum. He writes,

> Digital expression allows an ease of movement across the perceiver spectrum that was impossible in two-dimensional fixed media of expression. If it is important that we learn to see the world through others' eyes, digital expression offers powerful new tools to do so. . . . All of the now myriad techniques for manipulating images and sounds . . . allow us to work at various places along the perceiver spectrum in new ways. (165–66)

These new techniques create perspectives readers and authors can explore. By reimagining the complex relationship between content and style, we open new ways to think about digital writing.

OurSpace Style

One of the primary changes the digital age brings to the writer's method is the logic of selection. Lev Manovich writes, in *The Language of New Media*,

> While previously the great text of culture from which the artist created her own unique "tissue of quotations" was bubbling and shimmering somewhere below consciousness, now it has become externalized. . . . The World Wide Web takes this process to the next level: it encourages the creation of texts that consist entirely of pointers to other texts that are already on the Web. One does not have to add any original writing; it is enough to select from what already exists. (127)

Whereas previous texts used selection subtly, new media texts overtly and explicitly weave together multiple media pieces. As Manovich suggests, this change makes explicit the process of selection as art and alters our understanding of artistic endeavor. But at the same time, he laments this new logic, suggesting that the only authentic response to overwhelming choice is minimalism. If new media demands choices of us, making such choices limits our expression to the author's options. The only response is to not choose: individual expression "can only be accomplished by refusing all options and customization, and, ultimately, by refusing all forms of interactivity" (129). But the changing shape of electronic authoring presents us with another option—Ugly.

One cause of paradigm shift from inspired genius to clever *bricoleur* is that art production tools and processes have become more widely available. Manovich's discussion about new media tools echoes Walter Benjamin's argument that mechanically reproduced art depreciates art auras and overturns conventional power structures by depreciating the aura around art. If mechanical reproduction disentangled art from aura, production tools must disentangle style from taste. Manovich suggests that making selections interpolates users into identity categories mapped out by the selections available; thus, making no choices refuses such interpolation. But refusing choices also defies the creative impulse and rejects the possibilities opened by new technologies. What if users amplified the logic of selection, rather than minimizing it? What if, instead of selecting from the choices offered, they brought a wider perspective to their art? If minimal selection represents one answer to Manovich's conundrum, hyperselection represents the other.

In his daily v-blog, *The Show,* Internet humorist zeFrank cultivates this idea in speaking about his "I knows me some ugly MySpace" contest:

> As people start learning and experimenting with these languages of authorship, they don't necessarily follow the rules of good taste. This scares the shit out of designers. In MySpace, millions of people have opted out of pre-made templates that "work" in exchange for ugly. Ugly when compared to pre-existing notions of taste is a bummer. But ugly as a representation of mass experimentation and learning is pretty damn cool. . . . Over time as consumer-created media engulfs the other kind, it's possible that completely new norms develop around the notions of talent and artistic ability. Happy Ugly.

The participants in zeFrank's contest work via the logic of selection, but they also eschewed both templates available in MySpace and the conventions of "good" design. In choosing their own visual cacophony, they blazed a trail around the paradox of selection that haunts Manovich.

These untrained designers highlight the power imbalance between arbiters of good taste—designers—and Web users. Ugly contestants select, in Manovich's sense, design elements and styles, but they use such elements in ways beyond the original creators' intent. This process of sampling and experimentation occurs at the level of code as well, where Web authors experiment with bits of code, whether or not they understand them. Never before have stylistic techniques been so readily copied. Amateur designers need not learn to draw perspective or learn how shadows work; they just need to know how to copy and paste code. Open standards for Web design pull back the curtain, giving design to everyone.

This democratization of style represents a continuing upset of print literacy's hierarchy and power. In the Web's early days, the production model still mimicked print in significant ways. For instance, *The Elements of Style* cautions the beginning writer, "Unless he is certain of doing as well [as skilled writers], he will probably do best to follow the rules. After he has learned, by their guidance, to write plain English adequate for everyday uses, let him look, for the secrets of style, to the study of the masters of literature" (Strunk et al. 6). Or, in common parlance, writers need to know the rules before they can break them. Early Web design posed similar obstacles, some maintained by convention, some by technology: authors needed obscure knowledge to code their pages, they needed to buy Web space, and they needed to understand acronyms such as FTP. But it wasn't long before Web authorship became anyone's domain, first via software such as Dreamweaver and FrontPage, and later through user-friendly authoring platforms such as blogs and wikis. These new publication modes upended traditional hierarchies of communication: now anyone with a blog can report the news, collaborate on public information documents, or post anything on MySpace. Such new environments demanded new concepts of communication and authority.

As tools for manipulating aesthetic elements become as available as textual tools, we must contemplate how such changes will alter our conception of style. Like the participants in zeFrank's contest, Web writers have begun styling their work. They work as designers, as collectors, as arrangers. For if acquiring the ability to control one's speech gives one power, so must the ability to control one's style. And writing in the digital age goes beyond alphabetic and verbal communication. Digital writing is style.

Fashion Style

Our culture's most practiced space for making style superordinate to content is fashion. In common parlance, fashion style represents a physical manifestation of one's inner self. It appears in our clothes, our attitude, our lifestyle; style is personality. To have style is to be both individual and hip.

Like the disarticulated approaches to style in writing instruction, some fashion experts suggest that style can be taught and marketed. Elsa Klensch, for instance, suggests a few simple strategies fashion-conscious women can wield to ensure that their style makes them more attractive, more popular, and more successful. At the same time, her rhetoric reinforces the notion that style represents the individual's personality. She writes, "Style is not about beauty, wealth, or even fashion. Style is rooted in a woman's knowing herself well enough to develop a consistent image, and then having the courage to project that image" (ix). Klensch constructs style as both rhetorical and authentic. In this way, she reiterates the form/content split: a woman's inner self reflects her content, and her best style does so with the utmost clarity and grace.

In subcultures, on the other hand, style opposes clear, normal communication. Dick Hebdige writes, in *Subculture: The Meaning of Style*,

> Style in subculture is, then, pregnant with significance. Its transformations go "against nature," interrupting the process of "normalization." As such, they are gestures, movements toward a speech which offends the "silent majority," which challenges the principle of unity and cohesion, which contradicts the myth of consensus. (18)

The style choices made by subculture groups speak against mainstream choices. Individuals use such choices to mark their membership in subculture groups or to communicate with others who share their semiotic domain. But these choices also actively oppose conventional style, causing "interference in the orderly sequence which leads from real events and phenomena to their representation in the media" (90). Although there still may be a sense of self invested in subcultural style choices, such choices mostly disrupt the orderly system established by linguistic and institutional hegemony.

This sense of style provides a different way to understand how non-linguistic elements of communication influence conveyed meanings. As opposed to the W3C approach, which separates meaning from the content it surrounds, subcultural style overturns the content/form hierarchy. Such alternatives inform our understanding of what style can be in the digital realm.

Web Style

With the rise of digital writing and the World Wide Web, we encounter style not as an adjective or a noun, but as a verb. More and more, style is something writers do, rather than something they have. In contrast with the traditional view that sees style, or design, as an ornament that adorns content, Web writing highlights the hybridity of these two channels of communication. From this perspective, the W3C's decision to separate style and content using the `<style>` tag and CSS represents a holding action against the tide erasing the content/form hierarchy. By contrast, scholars of new media and digital writing recognize the changing relationship between content and style. Gunther Kress explains, "We have moved from literacy as an enterprise founded on language to text-making as a matter of design. . . . From competence in use we have moved to competence in design and, with that, innovation and creativity (through the use of many modes) are now in the centre" (*Literacy* 105). For Kress, design has become an essential part of writing, inseparable from content.

Digital culture observes a similar process as Web writing evolves. Because early models of digital writing were structurally similar to print, the print conventions of content over form held strong. The standards and practices used in Web production continued to uphold these values. As the tools for individuals to produce Web texts have become more available, however, the conventions established by institutions of power have become less stable. Propriety in all things has destabilized, in style particularly. And it is in impropriety that Web writing grows most strongly: the improper mixing of the personal and the professional, the unwise focus on design over substance, the forbidden mix of form and content. Such experiences demand new kinds of Web reading, pushing us to uncover resources formerly depreciated under print literacy. We learn to read and write like flaneurs, guided by both reason and aesthetics. We write both content and style.

Works Cited

Berners-Lee, Tim, with Mark Fischetti. *Weaving the Web: The Original Design and Ultimate Destiny of the World Wide Web by Its Inventor.* San Francisco: Harper San Francisco, 1999.

Booth, Wayne C., Gregory G. Colomb, and Joseph M. Williams. *The Craft of Research.* Chicago: University of Chicago Press, 1995.

Corbett, Edward, and Sheryl Finkle. *The Little English Handbook.* 7th ed. New York: Watson-Guptill, 1995.

French, Christopher. *Associated Press Stylebook and Libel Manual.* New York: Laurel, 1990.

Hebdige, Dick. *Subculture: The Meaning of Style.* New York: Routledge, 1979.

Klensch, Elsa, with Beryl Meyer. *Style.* New York: Perigee, 1995.

Kress, Gunther. "'English' at the Crossroads: Rethinking Curricula of Communication in the Context of the Turn to the Visual." In *Passions, Pedagogies, and 21st Century Technologies,* edited by Gail E. Hawisher and Cynthia L. Selfe, 66–88. Logan: Utah State University Press, 1999.

———. *Literacy in the New Media Age.* New York: Routledge, 2003.

Lanham, Richard A. *The Economics of Attention: Style and Substance in the Age of Information.* Chicago: University of Chicago Press, 2006.

Levy, Steven. *Hackers: Heroes of the Computer Revolution.* Updated ed. New York: Penguin, 2001.

Lie, Håkon Wium. "Why Doesn't MSN Work with Opera?" *howcome.* 20 February 2003. Accessed 28 February 2007. http://people.opera.com/howcome/ .

Lunsford, Andrea. *The Everyday Writer.* 3rd ed. New York: Bedford/St. Martin's, 2004.

Maimon, Elaine P., and Janice H. Peritz. *A Writer's Resource.* New York: McGraw Hill, 2003.

Manovich, Lev. *The Language of New Media.* Cambridge, Mass.: MIT Press, 2002.

McLuhan, Marshall. *Understanding Media: The Extensions of Man.* Cambridge, Mass.: MIT Press, 1964.

Mischook, Stefan. "The Do's and Don'ts of Web Design." *KillerSites.com: Web Design Resources.* Accessed 5 August 2010. http://www.killersites.com/articles/articles_dosAndDontsWebDesign.htm.

Nielsen, Jakob. *Designing Web Usability: The Practice of Simplicity.* Berkeley, Calif.: Peachpit Press, 1999.

Norman, Donald. *The Design of Everyday Things.* New York: Currency, 1990.

Raymond, Eric. *Eric S. Raymond's Home Page*. Accessed 28 December 2008. http://www
.catb.org/~esr/.

Strunk, William, Jr., et al. *The Elements of Style*. 4th ed. New York: Longman, 1999.

Web Standards Project (WaSP). "Mission Statement." *Web Standards Project*. Accessed
25 February 2007. http://www.webstandards.org/.

Williams, Joseph M. *Style: Ten Lessons in Clarity and Grace*. 5th ed. New York: Longman,
1997.

W3Schools. "CSS Introduction." *CSS Tutorial*. Accessed 12 October 2006. http://www
.w3schools.com/.

zeFrank [Hosea Jan Frank]. "07–14–06." *The Show*. 14 July 2006. Accessed 1 December 2006.
http://www.zefrank.com/theshow/.

6   AN ACCIDENTAL IMPERATIVE

The Menacing Presence of

BRIAN WILLEMS

  is an HTML entity used to create a nonbreaking space.   *is a response to the inability of early HTML to match the layout demands users expected from word processing and related software. Indention, the creation of extra space between words, and the design of tables with empty cells were all applications of* . *The use of*   *offered an alternative to the single-pixel GIF file many Web writers used in order to artificially create extra space. However, improvements to HTML and the addition of Cascading Style Sheet selectors that permit alignment and positioning have made most of these uses irrelevant. Today, indenting and spatial control within sentences is far more fine grained. All in all, although*   *is a typographic carryover, its continued use addresses the difficulties Web writers still face in recreating the spacing found in print documents.*

THE PURPOSE OF A NONBREAKING SPACE is to create a blank space between two characters that cannot be separated by a line break. The reason for using a nonbreaking space is to deny text wrapping between characters. Although this is an example of the standard use of the nonbreaking space,   is also often seen as being overused by nonprofessional Web designers as a means of quickly getting blank space onto a Web page. An example of the more standard use of the nonbreaking space is in French, where a question mark is separated from the rest of a sentence by a space, whereas in English, it rests against the last character. If a spacebar-generated space, for instance, is used to separate the question mark from the rest of the sentence in an HTML document, some word-wrapping algorithms will allow the question mark to be put on the next line. If the nonbreaking space is used instead, this can be guaranteed not to happen (Gillam 443). The role of   on a Web page is therefore to separate two characters by a blank space, as well as to hold these two characters together by not allowing them to be separated by word wrapping (Heslop 47). In this sense, such characters, or bodies, can be thought of as those that are both separated and joined. Here, I use the dual nature of   to explore what it is that can be encoded in a tag or marker by examining a number of manifestations of the copresencing of unity and disjointedness in writing. This is developed through punctuation, futuristic bar codes, and pattern recognition.

What Is Punctuation, if Not Code?

Another way of envisioning a body as an entity that is both separated and joined by code is indicated by simply making this passive statement active by saying that code concomitantly separates and joins bodies. The reason for making what might seem like a tautological move here is to call attention to the imperative role of the nonbreaking space as markup that commands, or demands, such a dual nature in a body. The role of such a demand in an imperative is, first, a demand to be understood. Philosopher Werner Hamacher writes that the imperative is "the demand that understanding must take place; the imperative of understanding itself" (81). This statement forms one of the keystones of this essay. The first half of Hamacher's statement indicates how the addresser of the imperative demands an understanding of that imperative by the addressee, who is expected to understand and accomplish the action demanded. Therefore, a space is created between the two entities of addresser and addressee, and this space is indicated by understanding. The second half of Hamacher's statement is a claim for essence, or essencing, in the form of the imperative as an act of understanding itself. In order to attempt a way of conceiving the second half of this statement, a step backward is necessary, by which I mean that something was skipped in this description of the two halves of Hamacher's statement. This skipped space is occupied by a semicolon. A semicolon is punctuation used to separate and conjoin two equal clauses or ideas in a sentence. Hamacher's midquotation punctuation both separates the two halves of his sentence into two discrete ideas and unites these two ideas into a single statement. This copresencing of separation and unification is the same as what is central to the nonbreaking aspect of .

The role of the nonbreaking space is that of an imperative, but one with two demands: to separate and to unify. Although the simultaneity of these two demands might at first seem antithetical, Hamacher argues that both unity and separation are essential to understanding because it is impossible for any one being to understand another fully. There is always some misinterpretation, always room for error: where there is interpretation, there is misinterpretation. Hamacher argues that the difficulty in understanding another fully is actually an essential part of understanding itself, for "the lack of understanding must also be understood as a demand and, therefore, as a stimulus to understanding." The demand for understanding creates a lack of understanding (the other's discourse is never fully understood), and this lack of understanding functions itself as a demand for understanding. This is the relationship between the semicolon and the second half of Hamacher's statement: "the demand that understanding must take place; the imperative of understanding itself." A lack of understanding unconceals the essence of the demand itself. This essence is seen as the dual nature of the semicolon in the domain of punctuation, and as the dual demand of   in the domain of HTML. In other words, as Hamacher says, "from the very point where there can be no understanding arises the obligation to understand" (82).

Hamacher's semicolon has helped to develop the concept of the demand of unity and separation in the nonbreaking space.[1] In the same way that nonunderstanding is written into understanding, it can also be said that mistakes are written into law itself: "By demanding the law, the law forbids the law" (90). In Web practice, this means that nonstandard uses of HTML code are sometimes necessary for designers to achieve intended effects. So mistakes (according to such bodies as the World Wide Web Consortium) become law, standards, or patterns of design.

As an illustration of such a coding of mistakes, I mentioned earlier that   is often seen as being overused by nonprofessional Web designers because it is often chosen in lieu of a graphical space holder or Cascading Style Sheets (CSS). In browsing through a number of coding guidebooks, references to the nonbreaking space as "abused" (Powell 88), "shady" (Towers 211), and "dirty" (Castro 66) are not hard to come by. "Better yet, use CSS," suggests Elizabeth Castro. Such language (not often connected with markup) indicates the nonbreaking space's intimate coexistence with lawlessness as a shady character: the dual nature of the nonbreaking space both contains the law and that which goes beyond it.

The above is an example of user "error" in relation to the nonbreaking space, but the nonbreaking space also seems to lend itself to another kind of accident—being misunderstood by e-mail clients, word processors, or other agents that mistakenly transform   to & —for example, Microsoft Word can produce spurious   characters when

"Save as HTML" is invoked. This can be seen in editors too: loading a page in Dreamweaver, for example, can show   not just as space but as code. This second kind of error is what I call the menacing aspect of the nonbreaking space. The word *menace* has its roots in the Indo-European **men,* "to project." This meaning can be see in the Latin *minae,* "threats," but with the original meaning of the projecting points of walls. In this sense,   gets projected beyond the underlying code and onto the computer screen in the form of coding and code reading errors (although it is by no means the only tag to do so). This menacing aspect of the non-breaking space has been pointed out in a recent essay by Myra Moses and Steven Katz called "The Phantom Machine":

> Someone receives this message: "Love?   are you kidding?" In the context of a troubled relationship, he interprets the   as "and no bull-shit please," which makes all too perfect sense in the context of the email, further damaging the relationship. In actuality,   is html code for "non-breaking space," but was inserted in the message (many times more than shown here) in linguistically significant and emotionally crucial places as above by the email application. In cases like this, the phantom machine becomes visible—its code transmigrating software from the realm of the unseen to the screen. (93)

In this example, the failure to correctly interpret the manifest content of the e-mail is actually the impetus to a different kind of understanding: that of the machine humming underneath the code. If the message had been transmitted perfectly, the machine would have been forgotten. In this sense,   functions as a Brechtian alienation effect, or *Verfremdungseffekt,* pulling back the curtains separating the audience from performers backstage, revealing the fiction-creating device that is integral to the medium of the play.

What have been defined above are two different kinds of accidents attached to the nonbreaking space, first its user-centered dirty use in lieu of more professional techniques, and second its machine-centered ability to slip through code-reading devices such as mail clients and word processors. So what has been developed so far is the dual role of the nonbreaking space as an imperative that both separates two characters and holds them together; this is a similar role to the punctuation mark of the semicolon as a demand that indicates a space for understanding itself in the form of a lack of understanding; and finally two kinds of accidents attached to the   tag itself. In the following section, futuristic bar codes, called SPIMES, will be used to think through how these multiple facets of   relate to the construction of the identity of a body.

That Which Is Not Itself

For a body, take a bottle of wine as an example. Currently the bar code on a bottle of wine can tell you its name, price, where it comes from, and maybe a few other details. This is actually quite limited. One of the key concepts of a recent book of Bruce Sterling's nonfiction, *Shaping Things,* is the SPIME, a neologism used to describe the next step of bar codes or RFID (radiofrequency identification), and it opens up the amount and kind of information a product is able to give about itself. What a SPIME could do to a bottle of wine would be to make much more information available about that product, and this information would continually change in a wiki-type format allowing users and enthusiasts, whom Sterling calls wranglers, to add much more to the bar code than even the manufacturer could ever know about the product. So before purchasing a bottle of wine, a potential buyer could know how much the laborers were paid to manufacture the bottle, how long it will take to recycle, what energy has to be expended for the recycling process, what particles of glass will rub off into the wine and enter the bloodstream, what the long-term effects of these particles are, and so on (Sterling 15–24).[2] This fluctuating collection of information would then be the bottle's identity:

> The key to the SPIME is identity. *A SPIME is, by definition, the protagonist of a documented process. It is an historical entity with an accessible precise trajectory through space and time.* . . . The object becomes an instantiation of identity. It's named, and it broadcasts its name, then it can be tracked. That's a SPIME. (77, 105)

The SPIME is not only a collection of information about the bottle, but it also performs the identity of the bottle in a changeable format that can be bought and sold. The SPIME, in a sense, is a saturated totality that has crossed the limits of itself, opening up its identity to users and others who want to be involved in its creation: "In a SPIME world, the model *is* the entity, and everyone knows it" (96). The SPIME is a gathering of flickering bits of information, the bringing of outside world knowledge into the inside of identity. The identity of Sterling's bottle is not the bottle itself but rather what is given to it, making the SPIME a storehouse for other people's changes.

This bringing of the outside in is what I believe Moses and Katz call "the phantom machine." In their example of an e-mail client carrying the nonbreaking space as code into the body of a potential love letter, the identity of the nonbreaking space becomes that of the machine that reads it. The dual nature of   allows for a space in which accidents can happen—accidents that allow for the markup's identity to be that which it itself is not, meaning here of the medium through which it passes.

In order to try and clarify this relationship of that which is not itself, the thought of philosopher Jean-Luc Nancy will be brought in. The vision of the bar code as a saturated totality located in the crossing of borders is a powerful illustration of one of the most important contributions of Nancy's thought, which has been his reevaluation of Martin Heidegger's concept of *Mitsein* in the sense of how being-with actually precedes being itself. For Nancy, "identity is given by the difference that is not *its own*" (*Birth* 33). Identity is that which is given in a relation to the other as *exscription*, a term Nancy uses to index an outside being taken into the interior, thereby creating a kind of openness, or nakedness to another, which is actually that being's identity. Exscription is when the "'outside'— wholly *exscribed within* the text—is the infinite withdrawal of meaning by which each existence exists" (339). Exscription is both that of the embodiment of another along with a form of withdrawal: a double identity such as that found in . It is the dissolving of a limit (the limit of one being and another) and of a kind of totality that takes the limit's place. In Sterling's language, it is the multifarious identity of the SPIME. Nancy develops his reading of being-with in the following passage:

> Limits of matter (gases, liquids, solids), limits of kingdoms (mineral, vegetable, animal), limits of the sexes, limits of bodies, limits where sense becomes impossible, absolutely exposed, poured out, removed from any mystery, offered as the infinitely folded and unfolded line of all the bodies that make up a world. This world is their exhibition, that is, also their risk. Bodies run the risk of resisting one another in an impenetrable fashion, but they also run the risk of meeting and dissolving into one another. This double risk comes down to the same thing: abolishing the limit, the touch, the absolute, becoming substance, becoming God, becoming the Subject of speculative subjectivity. This is no longer the ab-solute, but saturated in totality. (206)

The relation between the abolition of a limit and becoming saturated totality is the dual nature of the identity of being-with, it is the accidental and imperative nature of the nonbreaking space, it is the identity of the

SPIME contained in the ever-changing input of its users. Christopher Fynsk, drawing on the thought of Nancy, develops this concept of being-with alongside a rethinking of Heidegger's *Mitsein:*

> Mitsein . . . cannot be understood in terms of the traditional notions of identification . . . this representative figure [for the German people to identify with] cannot be *like* anyone or anything else, and it is a *Gleichnis* of the holy only if we abandon any definition of this term elaborated in the aesthetic tradition. The poet figures only his receptivity to the holy and merely announces what, in coming, appears as absent. (*Heidegger* 218)

Fynsk reads the concept of a holy totality together with the possibility of dissolution. Both are inherent in a naked being-with. This is the reason that the emptiness of the nonbreaking space, or of Hamacher's semicolon, is important here. The impetus for understanding is created in an absence that is aporetic in its being both the blank space on a Web page and the location of an imperative, a dual demand both of dissolution and totality. Sterling's SPIME indicates its identity in a similar manner, as nothing but relationship. Nancy develops this concept of being-with in *Being Singular Plural:* "There is no 'meaning' except by virtue of a 'self,' of some form or another. . . . But there is no 'self' except by virtue of a 'with,' which, in fact, structures it. This would have to be the axiom of any analytic that is to be called coexistential" (94). The nonbreaking space is a way of looking at identity that has its multiple nature built into its sense of self; it is an example of naked emptiness promoting identity and action. This is the connection between what Fynsk reads as the absence contained in being-with and Nancy's concept of a holy brand of totality. The nonbreaking space is an instance of the coexistential because it is multiple, motile, and fresh. It is a vision of a singular instance of plurality.

But before there is too much excitement, a warning from Max Horkheimer and Theodor Adorno should be kept in mind here. Horkheimer and Adorno, in the opening pages of *Dialectic of Enlightenment,* warn against the feeling of understanding and accomplishment that can be a part of exscripting the voice of the other. Such exscription can first be seen as a relation to the divine: "If the tree is addressed no longer as simply a tree but as evidence of something else, a location of *mana,* language expresses the contradiction that is at the same time itself and something other than itself, identical and not identical." However, a move away from the divine and toward enlightenment is dangerous because although the same nonidentity is present, it is hidden under a comforting veil of understanding: "Humans believe themselves free of fear when there is no longer anything unknown. This has determined the path of demythologization, of enlightenment, which equates the living with the nonliving as myth had equated the nonliving with the living. Enlightenment is mythical fear radicalized." The examples that Sterling uses as what can be incorporated into the SPIME are unknown things commonly feared: health risks, child labor, unfair work practices, and environmental issues. According to Sterling, making these things known will change people's behavior. The lesson from Horkheimer and Adorno's reading of the Enlightenment is that such an incorporation of the known should not be confused with control, for it is born out of terror: "it emerges from the cry of terror, which is the doubling, the mere tautology of terror itself. . . . Nothing is allowed to remain outside, since the mere idea of the 'outside' is the real source of fear" (11).

In order to understand how the coordinates of   are being developed here away from the kind of totalization of fear that Horkheimer and Adorno warn against, a brief look at another concept of Sterling's, the biot, may be helpful.

The biot is important because it is "neither an object nor a person"; it "would be the logical intermeshing, the blurring of the boundary between Wrangler and SPIME" (Sterling 134). The biot is a way to hold back the reification of understanding. As humans, we eat, grow, and excrete flows of energy. In this metabolic process, we interact with the environment, our cells absorbing both nutrients and pollutants. At the moment, these processes take place below the radar of our awareness, meaning that our identity is not usually disrupted by their taking place. But the biot would be a being that would not only be aware of these changes, but would also be able to design processes that could manipulate them (135). Once this happens, "the ultimate consumer item is the Consumer" (137). The outside has become inside in the form of a being influencing its own evolution in an action of exscription. This is self-reflexive in the sense that the consumer is both product and producer. A line has been crossed here—a line that indicated outside and inside. The biot is an example of self-reflexive exscription, as a body that can touch itself, that can, in the words of Hamacher, understand itself. However, this understanding is not closed to its opposites—misunderstanding and error. In fact, as developed in the next section, it wears error on its sleeve, just like the menacing nonbreaking space.

Pattern Recognized

It is important to remember that the nonbreaking space functions within the electronic domain of pattern and noise. The interaction between pattern and noise comes out of the computer science version of pattern recognition. The reason pattern recognition is being looked at is that it is a vibrant domain of both inclusion and separation. It is also the location of the misreadings of   in e-mail clients. For example, when the code slips into content, it is a signal that it has not been recognized as belonging to the pattern of its intended category. In this way,   is that which might disrupt notions of how totalized pattern recognition is itself.

When functioning in a digital manner, pattern recognition can be used for computers reading the world as either on or off (a 0 or 1) so that they can make judgments in a similar manner as human beings, who have highly developed pattern recognition skills. In this way, computers can work along the lines of the not so black-and-white choices made by the human brain. Pattern recognition can be seen as a three-step process that assumes the following:

> (1) that complex visual or auditory stimuli are represented perceptually as points in a multidimensional, geometric space; (2) that similarity judgments are (inversely) related to interstimulus distance in the perceptual space; and (3) that identification judgments are described by a probabilistic decision rule based on the pattern of interpoint distances in the perceptual space. (Getty, Swets, and Swets 161)

Chris Bishop, in *Neural Networks for Pattern Recognition,* begins by defining pattern recognition with the frequently used example of how a computer might recognize the difference between two handwritten letters, such as *a* and *b.* Pattern recognition enables a computer to recognize the differences between these letters even when there exist, each time a person writes one down, many differences in handwriting and penmanship from one being to another (2). One way to think of this is to imagine a grid placed over each letter. The computer would contain an algorithm that could decide whether each square of the grid was closer to empty or full (and hence a 0 or 1). It would then be possible to have the computer look at each decision on a larger scale and decide whether the pattern fit within the established parameters for a letter. If a pattern does not fit in with either *a* or *b* (or any other sign it is programmed to recognize), then that input is still there, but it is just noise, the absence of pattern, non-information to the machine.[3] This is the menacing that allows   to bring forth the phantom machine.

Although this description offers a basic introduction to the digital processing of pattern recognition, the human brain works according to a different structure, one that is extraordinarily parallel. Ray Kurzweil describes the relationship between biological and nonbiological pattern recognition in this way:

> Human skills are able to develop only in ways that have been evolutionarily encouraged. Those skills, which are primarily based on massively parallel pattern recognition, provide proficiency for certain tasks, such as distinguishing faces, identifying objects, and recognizing language sounds. But they're not suited for many others, such as determining patterns in financial data. Once we fully master pattern-recognition paradigms, machine methods can apply these techniques to any type of pattern. (260–61)

Nondigital means for a computational device to recognize patterns include Bayesian nets, Markov modeling, and neural nets. The last can be taken as an example of how mechanical pattern recognition is moving away from the digital domain and closer to current understandings of the methods of the human brain. The reason that this is important in looking at   is that these nondigital methods are self-reflexive and error incorporating. What is at stake is that the medium of analog pattern recognition is a path that welcomes the accidental imperative to travel from code into writing.

The basis of neural nets is that they begin in ignorance. The input is broken down into a number of points (pixels in an image, for example) and each of these points is awarded, indiscriminately, a synoptic strength that indicates its importance. The output of each point is arbitrarily connected to other points on different levels of neurons. Once the activity of these connections exceeds a determined level, a random answer is generated. Neural nets then depend on guidance regarding the correctness of their answers. Connections formed with answers deemed correct by a teacher are strengthened, while those formed during incorrect procedures are not. Kurzweil continues, "Experiments have shown that neural nets can learn their subject matter even with unreliable teachers. If the teacher is correct only 60 percent of the time, the student neural net will still learn its lessons" (269).

The neural net is a form of mechanical pattern recognition in which error is not only present, but essential. Such an immanent demand for understanding born out of nonunderstanding is a neurological event, and as such seems to belong to the well-known neurological studies from Benjamin Libet. These studies suggest that the neural activity required to take an action occurs about a third of a second before the decision has been made in the brain, therefore suggesting that there is less of a totalizing command of ourselves at our fingertips than previously thought. Although these studies have raised numerous debates regarding the design of human freedom, philosopher Slavoj Žižek sees the accident as playing a pivotal role in Libet's experiments:

> What if failure comes first, what if the "subject" is nothing but the void, the gap, opened by the failure of reflection? What if all the figures of positive self-acquaintance are just so many secondary "fillers" of this primordial gap? Every recognition of the subject in an image or a signifying trait (in short: every identification) already betrays its core; every jubilant "That's me!" already contains the seed of "That's not me!" However, what if, far from consisting in some substantial kernel of identity, inaccessible to reflective recuperation, the subject (as distinct from substance) emerges in this very moment of the failure of identification? (244)

Žižek's location of the self is not in a totalizing identification but rather exactly in its lack. Another way to say this is that the role of the accident is essential in understanding. Žižek has shown that the exscription of misunderstanding into understanding, of that of error into code, is the only possible way to approach understanding at all.

The neural net model of pattern recognition incorporates a learning curve; it is an example of the exscribing of law within the lawless. Hamacher's kind of thinking is being taken at face value here; misunderstanding is seen as the only impetus to understanding in the neural net. The lawful and lawless are developed by Kurzweil in this way:

> Order is not the same as the opposite of disorder. If disorder represents a random sequence of events, the opposite of disorder should be "not randomness." Information is a sequence of data that is meaningful in a process, such as the DNA code of an organism or the bits in a computer program. "Noise," on the other hand, is a random sequence. (38)

This relationship between pattern recognition and noise has also been thought by N. Katherine Hayles, who, in exploring the change from a thinking of presence/absence to that of pattern/noise, says: "It is a pattern rather than a presence, defined by the probability distribution of the coding elements composing the message. If information is pattern, then non-information should be the absence of pattern, that is, randomness" (25). The change from presence to pattern has led Hayles to develop Lacan's concept of the floating signifier into one that flickers, meaning that at a keystroke, an entire entity can change in a global fashion—for example, by changing the font throughout a whole text document. The reason I look at Hayles here is that she brings the concepts of pattern and noise into the world of writing. For Hayles, flickering signifiers are "characterized by their tendency toward unexpected metamorphoses, attenuations, and dispersions" (30). This is the reason that parallel-processing pattern recognition programs such as neural nets are being examined here: the flicker, much like genetic mutation or genetic drift, offers hundreds of useless solutions and a few stunning ones. Error is essential to the identity of the system.

It is from within this electric flickering that the accidental imperative of   functions:

> As the emphasis shifts to pattern and randomness, characteristics of print texts that used to be transparent (because they were so pervasive) are becoming visible again through their differences from digital textuality. We lose the opportunity to understand the implications of these shifts if we mistake the dominance of pattern/randomness for the disappearance of the material world. In fact, it is precisely because material interfaces have changed that pattern and randomness can be perceived as dominant over presence and absence. The pattern/randomness dialectic does not erase the material world; information in fact derives its efficacy from the material infrastructures it appears to obscure. (Hayles 28)

This is the projecting/menacing aspect of   that was discussed in relation to the comment from Moses and Katz on "code transmigrating software from the realm of the unseen to the screen" in the form of a misunderstood love letter. The flickering signifier is that which gains its impetus toward understanding from the accidents inherent in the system it inhabits: it "derives its efficacy from the material infrastructures it appears to obscure" and makes that infrastructure become meaning, like Hamacher's semicolon. Hayles can therefore be said to take an immanent view of the creation of the new in the world. By this I mean that a way to understanding something beyond the everyday is inherent in the everyday itself, embodied within it, as shown by Libet and Žižek. This is the bifurcation between patterns and the imperative of Hamacher. A way of reaching beyond what is is not to look away but instead to allow the misunderstanding inherent in all attempts at understanding to be brought forth; that which is, by incorporating that which it is not in a self-reflexive fashion, becomes understanding itself.

As a way of conclusion, I will use an example from the domain of literature to illustrate the flickering imperative of   as it appears on the page. *Pattern Recognition* is the title of a novel by William Gibson in which a woman named Cayce is trying to make sense of something called "the footage." The footage is a number of untraceable scenes that are appearing on the Internet in a seemingly random order. There is great speculation on whether the images are found or made, filmed in sequence and then chopped up and uploaded out of order, or filmed out of sequence. There is also a debate regarding whether there is a planned order for the pieces or not. There are a number of nonofficial cuts of the footage put together, fan sites, and discussion boards. Cayce is sent by a wealthy patron to discover the creator of the footage. Although the pattern, or lack thereof, for the images is widely debated, a clue is eventually found. A number is found encrypted into one of the pieces of footage. The information is located in the white noise in the seemingly empty space of a scene in the footage. From a discussion on the main fan message board in the novel:

> Remember the whiteout, when they kiss? As though something explodes, overhead? . . . Blank screen. Taki says that "Mystic" decrypted this graphic [the sequence code] from that whiteness. As to how blankness can yield image, I do not pretend to know, though I suppose that is the question, ultimately, that underlies the entire history of art. (170)

The code for putting the film together is set within the whiteout of a kiss, "this image, extracted from that white flare" (173). From nothing, a whiteout, comes information, or meaning. The whiteout holds a similar place to the blankness on the page that is the result of using , as well as the contentless gesture of the semicolon. Understanding of the footage is born out of nonunderstanding; the emptiness of the whiteout is actually the secret the footage's junkies have been waiting for. Gibson's novel foregrounds the copresencing of emptiness and fullness. It throws such a conundrum up onto the screen, which can also be seen in the name of the main character.[4]

Such a flickering conundrum can be found in the way Cayce herself fits into the larger pattern of Gibson's work: Cayce's name is similar to name of the male protagonist in Gibson's first novel, *Neuromancer:* Case. The tension and difference between the two names is not ignored but rather brought forth: it menaces the novel. Toward the beginning of the book, Cayce meets a collector of obsolete technology:

> "My name is Voytek Biroshak."
>
> "Call me Ishmael," she [Cayce] says, walking on.
>
> "A girl's name?" Eager and doglike beside her. Some species of weird nerd innocence that somehow she accepts.
>
> "No. It's Cayce."
>
> "Case?"
>
> "Actually," she finds herself explaining, "it should be pronounced 'Casey,' like the last name of the man my mother named me after. But I don't." (31)

Cayce recognized her own participation in the eruption of nonrecognition. This eruption takes precedence in the code switching from speech to the written page because the difference between phonemes and morphemes only becomes apparent in a moment of *différance* when the pronunciation of Cayce's name is spelled out. In this single moment of reference, the phantom machine has been expressed by the copresencing of understanding and misunderstanding that is inherent in the menacing imperative of , which can be formulated by using Žižek's words, quoted above: "every jubilant 'That's me!' already contains the seed of 'That's not me!'" The "That's me!" of Cayce already contains the "That's not me!" of Case. The law contains what is beyond the law. As Cayce is an instance of sheltering this difference in a novel, the nonbreaking space is an instance of this difference in the domain of HTML. Gibson gives an example of this self-referential exscription in the character of Cayce, who incorporates error into her renaming; she names herself as that which is not herself. Cayce has become code.

Notes

1. In a similar manner to the discussion of punctuation in Hamacher, a colon from an essay title by Martin Heidegger has received much attention. The Heideggerian colon in question is from the title *"Das Wesen der Sprache—: Die Sprache des Wesens"* ["The essence of language—: The language of essence"] (Heidegger 72). The role of the colon in this phrase has been taken up by Avital Ronell, who says that "for Heidegger, even the colon can place a call. In fact, the call is made at a point of disconnection that cites and recites itself in the cleavage between word and thing" (167). By call, Ronell means a call

both from and to the outside of the everyday, here coming from within the code of language itself in the form of punctuation. Christopher Fynsk says that "the colon, in fact, marks the point where the concept begins to lose its grasp and where *Wesen* [essence] and *Sprache* [language] begin to undergo a radical transformation of meaning. But the speaking of the phrase in full unfolding and 'hinting' can only be characterized as that of thought" (*Language* 41). These readings of Heidegger's colon emphasize the transformative power of thought as punctuation in a similar way to Hamacher's semicolon, which creates a space for an attempt at the essencing of understanding.

2. It almost seems as if Sterling has taken the following passage from Karl Marx's *Capital* as a direct challenge: "As the taste of the porridge does not tell you who grew the oats, no more does the simple process tell you of itself what are the social conditions under which it is taking place, whether under the slave-owner's brutal lash, or the anxious eye of the capitalist, whether Cincinnatus carries it on in tilling his modest farm or a savage in killing wild animals with stones" (184). Regarding the validity of the SPIME as a concept, its entry has been deleted from Wikipedia because of supposed non-third-party interest, and the concept has been demoted to a neologism. See http://en.wikipedia.org/wiki/Wikipedia:Articles_for_deletion/Spime. Sterling himself keeps up a SPIME watch on his blog, "Beyond the Beyond," at *Wired*: http://blog.wired.com/sterling/.

3. Kenneth Sayre, in his classic 1965 book on pattern recognition, defines the difference between pattern and nonpattern: "The difference between the patterned and the nonpatterned group in each of these cases is not a matter of order or lack of order. Each group in these examples is ordered, for each member of the group can be related to other members within a common frame of reference. The difference rather has to do with the types of order displayed among the members of the group. Stated quite imprecisely, the difference is this: given an arrangement of a subset (some subset) of the group of elements, something can be anticipated regarding the arrangement of the remainder of the group. No subset of elements of a nonpatterned group, however, provides a clue regarding the arrangement of the remaining elements within the group. In other words, although the elements of a nonpatterned group can be arranged in some unambiguous order, the ordering is essentially random, while in a patterned group knowledge of the arrangement of a subset provides a basis upon which the arrangement of the remainder of the group might be predicted" (150–51).

4. For a reading of how the external footage is actually a product of a U.S. Army M18A1 Claymore mine embedded within the creator's skull, see my essay "Devoted to Fake."

Works Cited

Bishop, Chris. *Neural Networks for Pattern Recognition*. Oxford: Oxford University Press, 1995.

Castro, Elizabeth. *HTML for the World Wide Web with XHTML and CSS: Visual QuickStart Guide*. Berkeley, Calif.: Peachpit Press, 2003.

Fynsk, Christopher. *Heidegger: Thought and Historicity*. Ithaca, N.Y.: Cornell University Press, 1993.

———. *Language and Relation: . . . That There Is Language*. Stanford, Calif.: Stanford University Press, 1996.

Getty, David, John Swets, and Joel Swets. "Multidimensional Perceptual Spaces: Similarity Judgment and Identification." In *Auditory and Visual Pattern Recognition*, edited by David Getty and James Howard Jr., 161–80. Hillsdale, N.J.: Lawrence Erlbaum Associates, 1981.

Gibson, William. *Pattern Recognition*. New York: Berkley Books, 2004.

Gillam, Richard. *Unicode Demystified: A Practical Programmer's Guide to the Encoding Standard*. Boston: Addison-Wesley, 2003.

Hamacher, Werner. "The Promise of Interpretation: Remarks on the Hermeneutic Imperative in Kant and Nietzsche." In *Premises: Essays on Philosophy and Literature from Kant to Celan*, translated by Peter Fenves, 81–142. Stanford, Calif.: Stanford University Press, 1996.

Hayles, N. Katherine. *How We Became Posthuman: Virtual Bodies in Cybernetics, Literature, and Informatics*. Chicago: University of Chicago Press, 1999.

Heidegger, Martin. *On the Way to Language*. Translated by Peter Hertz. New York: Harper and Row, 1971.

Heslop, Brent. *HTML Publishing and the Internet*. Raleigh, N.C.: Ventana Communications Group, 1998.

Horkheimer, Max, and Theodor Adorno. *Dialectic of Enlightenment: Philosophical Fragments*. Translated by Edmund Jephcott. Stanford, Calif.: Stanford University Press: 2002.

Kurzweil, Ray. *The Singularity Is Near: When Humans Transcend Biology*. New York: Penguin, 2005.

Libet, Benjamin. "Subjective and Neuronal Time Factors in Conscious Sensory Experience, Studied in Man, and Their Implications for the Mind–Brain Relationship." In *Mind and Brain: The Many-Faceted Problems*, edited by John Eccles, 185–88. New York: Paragon House, 1990.

Marx, Karl. *Capital: Vol. 1*. Translated by Samuel Moore and Edward Aveling. New York: International Publishers, 1967.

Moses, Myra, and Steven Katz. "The Phantom Machine: The Invisible Ideology of Email (A Cultural Critique)." In *Critical Power Tools: Technical Communication and Cultural Studies*, edited by J. Blake Scott, Bernadette Longo, and Katherine V. Willis, 71–105. Albany: SUNY Press, 2006.

Nancy, Jean-Luc. *Being Singular Plural*. Translated by Robert Richardson and Anne O'Byrne. Stanford, Calif.: Stanford University Press, 2000.

———. *The Birth to Presence*. Translated by Brian Holmes et al. Stanford, Calif.: Stanford University Press, 1993.

Powell, Thomas. *HTML and XHTML: The Complete Reference*. Emeryville, Calif.: McGraw-Hill/Osborne, 2003.

Ronell, Avital. *The Telephone Book: Technology, Schizophrenia, Electric Speech*. Lincoln: University of Nebraska Press, 1989.

Sayre, Kenneth. *Recognition: A Study in the Philosophy of Artificial Intelligence*. Notre Dame, Ind.: University of Notre Dame Press, 1965.

Sterling, Bruce. *Shaping Things*. Cambridge, Mass.: MIT Press, 2005.

Towers, J. Tarin. *Dreamweaver 4 for Windows and Macintosh: Visual QuickStart Guide*. Berkeley, Calif.: Peachpit Press, 2001.

Willems, Brian. "Devoted to Fake." *electronic book review*. 26 March 2008. Accessed 18 November 2008. http://www.electronicbookreview.com/.

Žižek, Slavoj. *The Parallax View*. Cambridge, Mass.: MIT Press, 2006.

7   THE EVIL TAGS, `<blink>` AND `<marquee>`

*Two Icons of Early HTML and Why Some People
Love to Hate Them*

BOB WHIPPLE

`<blink>` is a Netscape proprietary tag, and `<marquee>` *was created by Microsoft
for its Internet Explorer browser. As elements of the famous browser wars, the two
tags have long been vilified, even labeled evil. The tags reflect the early influence of
advertising on the Web:* `<blink>` *flickers text like a neon sign;* `<marquee>` *runs
text along the bottom of a Web page like a ticker-tape message. Webzine pioneer Suck
blasted* `<blink>` *as a "micro-ad," warning, "Soundtracks, exotic buttoneering, inline
video, and of course, even more absurdly pervasive animation flash is creeping up on
a terminal near you."* `<marquee>` *didn't fare much better. Jakob Nielsen wrote, "A
web page should not emulate Times Square in New York City in its constant attack
on the human senses: give your user some peace and quiet to actually read the text!"
Neither tag is used with frequency today.*

IMAGINE, IF YOU WILL, a Web page author in 1995. Armed with a working knowledge of HTML 2.0, a text editor, and a computer, he crafts his
first Web page. Making mistakes, correcting them, forgetting elements,
replacing them, he plugs on, determined to make a page that works!

He finishes the file, names it, copies to his rudimentary server—a speedy
486—breathlessly trots to a computer five feet away, loads the early-
generation browser, and . . . eureka! There in front of his eyes is a Times
New Roman masterpiece, innocent of any knowledge of Web page design,
its white background contrasting starkly with the jet-black font, and, at
the bottom of the page, a moving phrase, creeping slowly from right to
left, little verbal ants treading inexorably across a picnic cloth:

> *Welcome . . . To . . . Our . . . Web . . . Page. . . .*

Or, flashing intermittently and continuously:

Howdy!
(Howdy!)
(Howdy!)
(Howdy!)

The great thing about HTML, of course, was the fact that text no longer behaved as text when the HTML page was viewed through a browser. No longer did text just sit there. Instead, with the (arguably, as we'll see below) creative use of tags, text moved. Text shook. Text jumped around. Text blinked. And text walked. In short, text did things that it had never done before. With HTML, the sameness and stasis of traditional text was now (more or less) easily augmented by the possibilities of graphics, pictures, and animations, not to mention the capability of hyperlinks. (Click—wait a moment—and shhhhwwoooom! You've been taken to another page! It's magic!)

And so went my introduction to multimediation, HTML, Web pages, Web servers—and the `<blink>` and `<marquee>` tags. As they did with me, these tags fascinated many new HTMLers creating their first Web pages. These elements, like most HTML, were simple to insert and simple to type, consisting, in their most basic forms, of a word with carets on either side. Both tags could be tuned more finely; `<blink>`, for example, could have additional specifiers governing the timing between blinks.

But, as with so many things, too much of a good thing is still too much. The perceived overuse of these tags soon after their invention caused irritation and denunciation in the early-HTML mid- to late 1990s, both as a result of their overall obtrusiveness and hard-to-miss-ness, as well as the fact that they just wouldn't stay still; they were viewed as the evil twins of Web design. The 3,000-year-long cultural memory associated with text made—and makes—readers feel more comfortable with words staying where they are on the page. Until very recently, we have expected our text, like perhaps our children, to stay still and be quiet. Fat chance. `<blink>` and `<marquee>` didn't do what they are expected to; indeed, they did what someone else—the Web author—told them to do, and what they seemed to do, especially in 1995, was flout conventions of what text should do—or, rather, not do. In effect, `<blink>` and `<marquee>` were the rebellious teenagers of HTML: always moving, never still, appearing where they're not expected, then disappearing,

But it's both too easy and incorrect to dismiss these two as simply evil twins. `<blink>` and `<marquee>`, among other tags, broke the fourth wall of textual consumption. Both created, and continue to create, effects both familiar and predictable, but which, when they were first introduced, were, in the context of most traditional text, quite novel. The story of these tags is the story of the introduction of movement and flash to text—not necessarily as in "flashing lights," or specifically Macromedia's Flash product (though Flash can cause flash), but as flashy, or attention getting (or demanding). The `<blink>` and `<marquee>` tags are, at least in their effect, uniquely American tags; certainly any stereotype one would care to name (American pushiness, need for attention, overbearing attitude, or flashiness) can be metaphorically compared with `<blink>`. And consider the multiple marquees of Broadway, of Times Square, of NASDAQ, ESPN, ABC News—or, for that matter, the scrolling ticker-tape messages running constantly at the bottom of the screen on Fox News, ESPN, or CNN; consider the luminescent extravaganzas of Las Vegas and its Sunset Strip. If America is not all about flashing, moving, blinking, bright lights, then flashing, blinking, bright lights nonetheless impart a uniquely twentieth-century American flavor. Flashing lights and ticker-tape-type marquees are part of what David Birdsell and Leo Groarke call "visual culture" (7)—more particularly, though, they are part of American visual culture. What follows in this essay will serve, then, as a starting point for situating these elements in this context.

Blinking and `<blink>`ing

Blinking and scrolling marquees are hardly new. Indeed, these tags mimic, and mirror, effects and phenomena that have existed for centuries. Humans have always been fascinated by flashing, blinking objects: the crackle of a fire, the twinkling of stars, the glimmer of a lightning bug or distant torch all make us look, wait, and wonder. Intermittent, blinking lights draw attention to themselves by the very fact that they are not constant—they change. The presence of something, then its absence, then its reappearance again naturally draws attention to a thing; the attention is gained by the disappearance, and then heightened or piqued by the reappearance. Anyone who has ever played peekaboo with a small child, hiding one's face, then revealing it, hiding, then revealing it, knows that the peals of the child's laughter are caused by the surprise and pleasure when the face reappears after it is covered up.

Blinking and moving electric lights are almost so much of a cultural commonplace in American culture as to resist easy notice; they are part of our background, part of the visual culture of contemporary American visual existence. According to Marshall McLuhan, "the electric light is pure information" (8). It is a communication medium as much as, say, the telephone. What it says, according to McLuhan, is the content of the medium (8). When the light blinks, it is doing something more than simply being an electric light; it is sending out a message—McLuhan's content. We can extend this, then, to argue that the blink is the medium—the vehicle, if you will—and that the text that it contains is what McLuhan calls the content.

This is not to say that the content is as easily apprehended in as absolute a way as we might like. Indeed, the example of the blinking light may well be as visible or explicit an example as we can get of Jacques Derrida's concept of *différance*, which he terms a "detour," or a "a delay," a kind of "temporizing" (8)—or, to put it another way, a putting off until a never-realized later. Blinking text both differs and defers, as Derrida dually defines *différance* (Johnson ix; Derrida 8). Blinking differs from traditional text, from the natural expectation that text will stay still. It also defers—indeed, it does so multiple times by delaying the reader's path through the reading; it also defers the meaning of the text by removing the text, however briefly, from sight. The constant motion gives the reader a sense of ongoingness, of expectant incompletion, of something there but not quite there, a sense reminiscent of Derrida's assertion that we can only get close to the meaning of a text, but never actually get to it; that the word cannot be what it represents, but a pointer to it, an approximation thereof: "When we cannot grasp or show the thing, state the present, the being-present, when the present cannot be presented, we signify, we go through the detour of the sign. We take or give signs. We signal. The sign, in this sense, is deferred presence" (9).

Blinking text, then, is a "signal" in the Derridean sense; it is a "substitution," a visual expression of this kind of "incomplete" or "deferred" (or "differing") presence (Derrida 9). The "incompletion" in the theoretical sense is ongoing; in a more literal sense, it may not seem so. After all, if we remain with the blinking text long enough, we can puzzle out its meaning.

As an HTML tag, `<blink>` simply causes the text within the nested `<blink> </blink>` tags to appear, disappear for a short interval, and then reappear, and so on, much the same as a blinking light turns on, then off, then on again, repeatedly (December and Ginsburg 1047). `<blink>` is supported in the Netscape, Firefox, and Camino browsers but not in the Safari and Internet Explorer browsers (as of February 2009). The `<blink>` tag's history—as well as the beginning of griping about the tag, it seems—

begins with the early days of the Netscape Corporation and its signature browser, Netscape Navigator. `<blink>` was developed by Netscape's Louis Montulli, an early developer of HTML standards, as well as the text-only Web browser Lynx (while a student at the University of Kansas); at Netscape, he developed several now-standard innovations, such as HTTP cookies and HTTP proxying ("Lou Montulli").

Most notably, though, Montulli is reported as having said in interviews that `<blink>` was "the worst thing I've ever done for the Internet" ("Blink Element"). And Montulli's reported opinion certainly is borne out by many other Web-based commentators. For example, the "HTML Hell Page" tells readers, "you know you're in design hell when you see . . . " and then lists "blinking text" at the top of a list undesirable Web design features, noting, "Blinking text makes it nearly impossible to pay attention to anything else on the page . . . if you abuse the blink tag, you deserve to be shot" (Raymond). Judith Wuseman asserts, "The most notorious nonstandard extension must be the Netscape blink tag which causes text to flash on and off to the annoyance of some Net users." And finally, Web design guru Jakob Nielsen reports: "Of course, `<blink>` is simply evil. Enough said" ("Original Top Ten"). Clearly, though the `<blink>` tag was popular in the early days of HTML (which I place from 1994 through 1998), it got little love from the growing ranks of HTML and Web design professionals. Although it's possible that their criticisms are a holdover from the aesthetics of static text and its expectations of readability, there is still something behind their vehemence—a sense that usability is hampered for the sake of flashiness.

Marquees and `<marquee>`s

The `<marquee>` tag finds its genesis in the classic 1940s and 1950s ticker-tape marquees providing scrolling messages, also seen currently in New York's Times Square and at the bottom of the screen on CNN, Fox News, and ESPN, among other networks. The original ticker-tape marquees achieved their effects with sequential lighting of incandescent lamps. Neurologically, the way that electric-light ticker-tape marquees act on the brain is fascinating. According to Lars Muckli et al.:

> While we know each bulb remains stationary, the lighting and dimming of each in succession makes it appear that light is moving across the marquee. Even when successive bulbs are separated by a large space, our brains fill in

the missing data to create the illusion that the motion has occurred smoothly from one point to the next.

Hence when we read a message on a ticker-tape marquee, we concentrate on what we see as a moving message, rather than on the fact that it is a sequence of static electric lights—or in the case of the `<marquee>` tag, pixels—flashing and dimming in a particular sequence.

It should be noted that though the `<marquee>` tag is characterized by motion, a marquee, in the physically built sense, does not have to move at all. A marquee is literally, according to the *Oxford English Dictionary*, "A canopy projecting over the main entrance to a building; spec. such a canopy at a theatre, cinema, etc., on which details of the entertainment or performers are displayed." The kind of marquee that gave birth to the tag, though, is what is often called a ticker-tape or scrolling marquee. Not surprisingly, the blinking light meets the marquee in the context of electric signs, which in many incarnations are nothing more than complex series of electric lights blinking in particular sequences. The famous Times Square news marquees of the mid-twentieth century, seen in countless films, generally date from the 1940s. According to Carl Wagner, CEO of Ohio's Wagner Electric Sign Company, unlike modern scrolling marquees, which are electronically or computer driven, these early marquees were mechanically driven, relying on differing cams to open and close the electrical contacts to create the illusion of motion in the messages. These scrolling signs are familiar to anyone conversant with American popular culture of the twentieth century. One can still find, perhaps in older surviving theaters in small town centers, or in downtown areas in larger cities, large marquees that—because there is only one screen, and therefore usually only one film showing—have enough space to show the film's title, actors, and even review blurbs, some of which are—or were—elaborately illuminated, sometimes with rows of lights around their borders, and sometimes on the inner surface of the projecting overhang, often neon, but often also incandescent lights, some of which might be blinking, often in a pattern in which a sequence of lights will chase each other around the marquee in a particular pattern. The modern marquees of Times Square, on the other hand, display the full panoply of colossal multimediation. No longer limited to ticker-tape presentation of text via incandescent bulb, they function much as gigantic Jumbotrons in sports stadiums, with scrolling text joined by video, animation, and sound.

The purpose of all of these marquees or marqueelike displays, whether text only or not, is, of course, much like that of blinking lights, blinking electric signs, and the `<blink>` tag: to attract attention. The average human loses interest in things that stay still pretty quickly. But things that move—especially when they are not expected to—excite our interest. The ticker-tape marquee draws the reader/viewer toward, and ultimately into, the theater, the venue, or in the case of modern Times Square marquees, the very experience itself. Originally serving theaters, marquees are themselves theatrical; like actors exaggerating their vocal volume, makeup, and gestures to be seen in the farthest row in the back of the theater, marquees impress, they call, they beckon. Ticker-tape marquees, with their messages, inform as well, achieving the dual purpose of catching attention and conveying a message—close enough, probably, to be classified as infotainment.

The `<marquee>` tag, a descendant of the ticker-tape marquee, moving (usually) right to left, is a Microsoft tag ("HTML"), introduced in early versions of Microsoft Internet Explorer. As of February 2009, `<marquee>` is supported by the Firefox, Opera, Safari, and Camino browsers. Text placed between the `<marquee>` `</marquee>` tags will scroll from right to left across the screen at the point in the Web page that the tags are inserted. Additional parameters can be specified in the tags to govern the speed of the scrolling, the background color of the marquee, and the direction the text message moves (December and Ginsburg 327, 1003–6).

It's worth noting, too, that, ironically, while the function of ticker-tape marquees was to inform as well as catch attention, `<marquee>` messages in HTML are if not more than, then at least as much about the movement (perhaps what Marshall McLuhan would refer to as the medium, because the literal function of a marquee is to move a message) than the message. Think of it this way: the original ticker-tape marquees originally scrolled because it was an efficient way to get a message across and displayed; enormous multimediated building-sized electronic displays did not exist, and therefore, to provide a headline or news item, the marquee had to scroll it from right to left across a space the width of a single line of text. But on the Web, there is no such technological or space limitation. Except for the decision that a certain text needs to be paid particular attention to, there's no other reason for a text not to be placed, say, in the body of the Web page—indeed, a text can be placed in the body of Web page somewhat more easily than placing it between `<marquee>` tags. Thus, to put something in `<marquee>` is to make a deliberate choice to draw attention to the text by the very fact that it's moving.

And, to be honest, the poor `<marquee>` tag is not much more beloved of Web designers than the `<blink>` tag. The "HTML Hell Page" lists `<marquee>` third in its list of eighteen annoying Web design features. Additionally, the Web site Writing HTML states, as the title of its Rule 17, "Don't Blink, Don't Marquee!" adding, "rely on compelling *content*, rather than *cheap* attention grabbers." And finally, a recent post on LiveJournal—which accepts tags in entries—is titled simply, "Kill the Marquee Tag."

Assumptions of Fixity and the Rhetoric of Vegas

I've established that blinking text and scrolling marquees attract attention (as well as criticism). But we need to delve more deeply into why this is so, as well as think about what exactly blinking text and scrolling marquees mean as visual actions as well as the words that they may portray. Let's face it—we don't usually expect the text we read to do anything other than just sit there, and we have some very good reasons for thinking so. Text, as we have known it for well over 3,000 years, has largely been inscribed or written or painted or inked upon another thing—paper, stone, wood, clay, wax tablets, metal. It's been essentially the overlaying or inlaying of a static mark or marks upon another relatively static medium. Physically, then, we have been enculturated to think of written text as static, fixed, and unchanging within the confines of the medium that contains it.

Let's think, for a moment, about different kinds of fixity, related, but with particular distinctions. Let's consider two to start—physical fixity, as noted above, wherein something is literally, physically, imprinted or engraved on an immovable medium, such as stone or paper, and cognitive fixity, which came after the innovations of Gutenberg and his contemporaries of the early printing press. The former preceded the latter, but the latter did not naturally flow from the former. It was not until the arrival of the printing press in the late 1400s that a dramatic shift in the nature of what we consider a standard, or fixed, text, occurred. Manuscript books, prone to the natural human error of human copyists, never really had a fixed set of text that was reproduced accurately from copy to copy; any error in a version could be copied and perpetuated as an error into a copy of that version, with, perhaps, additional errors by the second copyist, and so on. Although text did not physically move, it incrementally, contextually, and substantively changed; there was no ability to more absolutely fix the letters on the page from one copy to the next. Inasmuch, for example, as we think of an edition of a book as a standardization of the text in a fixed manner, "the term 'edition' comes close to being an anachronism when applied to copies of a manuscript book" (Eisenstein 9). Printing changed this rapidly. Elizabeth Eisenstein tells us that printing shops grew into networks of publishing operations with editors and proofreaders, generating the ability to exactly reproduce text much faster than a medieval copyist could dream of doing. This exact reproducibility was taken for granted by readers, who saw text as unchanging

Thus, when the World Wide Web, via graphical browsers such as NCSA Mosaic and early Netscape came along, the motive effects of early HTML effectively shook up the dynamic of written textual production. The evil tags `<blink>` and `<marquee>` could hardly help but catch the fancy of whoever read the Web pages that used them. Indeed, let us consider Web pages—as many still do—as rather booklike, kind-of-print things that presented text that looks normal. (After all, we read them, and they're called pages, right?) One can imagine, then, the reaction when the text took off and started doing its little dance:

"Wait—what was that? It BLINKED! It's there, look, look—there it goes again—It's blinking! Look at that! (What's it saying?) Say—that's kinda cool—how can I make my Web page do that?"

"No, WAIT! Over there—it's MOVING! Wow—that's *cool!* How did you make it do that? What's that? Ok, bracket-marquee-bracket, then the message—I'll figure that out later—then bracket slash marquee bracket. OK. Got it."

And so it went.

But why should we really be surprised at the appeal (and subsequent apparent revulsion) of motion in text, of blinking and sliding from right to left? As noted earlier, blinking lights are constants in American visual culture; so, to a lesser extent, are scrolling marquees. American visual culture since the 1950s has been heavily influenced by what Robert Venturi, Denise Scott Brown, and Steven Izenour refer to as the "commercial vernacular" (8) in architecture. Indeed, in *Learning from Las Vegas: The Forgotten Symbolism of Architectural Form,* they provide an analysis of an inherent unity in the unique architectural elements of the Las Vegas Strip, with its gaudiness and overwhelming display, that can lead us to an understanding of how natural, predictable, and American such tags might be.

Venturi, Brown, and Izenour refer to the built icons of display and show in as Vegas as the "architecture of persuasion" (9). They note that "the [Las Vegas] Strip is all signs" (9) and that in such a context, "The sign is more important than the architecture" to which it is connected or to which it may point or lead (13). These constantly moving, flashing, and acting signs "inflect toward the highway" and "[leap] to connect the driver to the store" (13, 51–52). The drivers on the street or highway are, of course, potential consumers, so the sign reaches for them, beckons to them, invites them in, even though what may lie inside the building hidden behind the sign remains, for the moment, unknown. Movement here is paramount—the multiple movements of each sign invite movement toward the sign itself on the part of drivers who are themselves moving, the ultimate goal of the sign being movement into the building, whose importance, until that point, may be secondary to that of the sign.

There is, however, method to the apparent madness. "The order of the Strip *includes*" (52), and furthermore:

> It is not an order dominated by the expert and made easy for the eye. The moving eye in the moving body must work to pick out and interpret a variety of changing, juxtaposed orders. . . . It is the unity that "maintains, but only just maintains, a control over the clashing elements which compose it. Chaos is very near; its nearness, but its avoidance, gives . . . force." (52–53, quoting August Heckscher's *The Public Happiness*)

We can here see in the assertions of Venturi, Brown, and Izenour a new way to look at <blink> and <marquee>. They may be vulgar (even they call the Las Vegas sign a "vulgar extravaganza"; 13), but it is vulgarity with a purpose. If they annoy, they are annoying with meaning behind them, their style—manifested in blinking and scrolling—trumping, or at least equal to, their substance. Apparent chaos, or near chaos, has its purposes too. And although many readers may not feel comfortable accepting "force" from their reading texts, perhaps this is because they—we—have yet to develop the habits and skills of reading that allow for motive text, using the architecture of persuasion, to be a part of our text-consuming experience.

By themselves, <blink> and <marquee>, being relatively limited in what they can individually do, may indeed tend to wear out their welcome after a time. The first couple of Web pages one sees with blinking text are intriguing; after a while, not unlike a surfeit of ice cream, one may silently (or not) say: "Enough, already." Nielsen, Web designer and usability advocate, eschews complexity in Web sites (*Designing Web Usability* 380), and, one would assume, a kind of barely controlled, Vegaslike, force-inducing chaos as well. Web page readers, according to Nielsen, are busy, haven't a lot of time, and scan ("How Users Read"); blinking text and marquees tend to interrupt scanning, stop readers, and grab their attention; this delay may seem irksome. A message in a `<marquee>` tag is limited to what can scroll across a screen in (generally) one line; thus, `<marquee>` messages are necessarily shorter. As a result, blinking text says, inevitably, "Look at me! I'm different! I'm important! I'm here! I'm not here! Look to see when I will next appear and disappear!" By themselves, and devoid of the massive multiple and somewhat unified thematic context of the Vegas Strip (or, perhaps, a more contemporary multiply mediated Web page in which the blinking is one of several unitary and coherently designed elements), `<blink>` and `<marquee>` may draw more attention to themselves than desired. These tags by themselves draw the reader's sole—and eventually critical—attention.

Thus any displeasure with `<blink>` and `<marquee>` falls not so much on the nature of the text, but on the nature of the action of the text. In the attempts of early Web writers—the users of these tags—to be new, they are often being faulted for being new, for not conforming to static textual expectations. It's as if text, finally having the temerity to do something other than rest quietly on the page, is being chastised for doing something other than sitting quietly and behaving itself. One is almost tempted to say, in defense of the tags, "Yeah, but at least they tried something new"— or at least new for the stasis that characterized written text.

Conclusion

Web writing allows, and perhaps demands, movement, or at the least a break with the physical and cognitive fixity of text. Text no longer just sits there, nor, with the ephemerity of Web space, can any webbed text, moving or still, be said to be absolutely fixed. Hypertext, as well as successive tools, such as Flash, which Nielsen also criticizes as disruptive to efficient Web reading ("Flash"), Java, animation, audio, and video have shattered readerly expectations and in the process have made `<blink>` and `<marquee>` seem quaint and archaic. We may now, in the early twenty-first century, have outgrown these two idiosyncratic and controversial tags that made our early webbed days (at least momentarily) exciting. But we should remember them—and perhaps remember that motion and flashiness are as originally native to Web writing as HTML itself. After all, the fact that the luminous giants on the Vegas Strip may be vulgar and tawdry doesn't necessarily mean that they are ineffective at doing what they do: drawing people in. If we can assume that the designers of Vegas's signs—or the multimedia marquees of Manhattan—know what they are doing, and that they plan their approach, research their audience, and craft their messages as carefully as a traditional writer, then why should the effects of `<blink>` and `<marquee>` be cast into the darkness without a chance to be used in effective and appropriate ways?

Indeed, while the `<blink>` HTML tag may be gone, blinking text is not. The Web site Blinking Text Live provides customizable blinking text via code generated on the site that users can then paste into their own Web pages. Blinking is not dead (though as noted earlier, hardly to the liking of many Web designers). Let's not forget the `<blink>` tag, or eschew marquees, then. The evil tags `<blink>` and `<marquee>`, for a moment, knocked us back on our heels and made us think about what text should—and could—do. And it's doubtful that writing will ever be the same as a result.

Works Cited

Birdsell, David S., and Leo Groarke. "Toward a Theory of Visual Argument." *Argumentation and Advocacy* 33 (1996): 1–10.

"Blink Element." *Wikipedia.* Accessed 15 March 2009. http://en.wikipedia.org/.

Blinking Text Live. Accessed 15 March 2009. http://www.blinkingtextlive.com/.

December, John, and Mark Ginsburg. *HTML 3.2 and CGI Unleashed.* Indianapolis: Sams Net, 1996.

Derrida, Jacques. *Margins of Philosophy.* Translated by Alan Bass. Chicago: University of Chicago Press, 1982.

"Don't Blink, Don't Marquee!" *Writing HTML.* Accessed 23 February 2009. http://www.mcli.dist.maricopa.edu/.

Eisenstein, Elizabeth. *The Printing Revolution in Early Modern Europe.* 2nd ed. Cambridge: Cambridge University Press, 2005.

"HTML." *WWW: Beyond the Basics.* Accessed 23 October 2006. http://ei.cs.vt.edu/book/.

Johnson, Barbara. "Translator's Introduction." In *Dissemination,* by Jacques Derrida, vii–xxxiii. Chicago: University of Chicago Press, 1981.

"Kill the Marquee Tag." *LiveJournal.* 4 June 2003. Accessed 22 October 2006. http://community.livejournal.com/suggestions/.

"Lou Montulli." *Wikipedia.* Accessed 25 October 2006. http://en.wikipedia.org/.

McLuhan, Marshall. *Understanding Media.* New York: McGraw-Hill, 1964.

Muckli, Lars, et al. "Now You Don't See It, Now You Do: Filling In Creates the Illusion of Motion." *PLoS Biology* 3, no. 8: e290. 19 July 2005. Accessed 15 March 2009. http://www.plosbiology.org.

Nielsen, Jakob. *Designing Web Usability: The Practice of Simplicity.* Indianapolis: New Riders, 2000.

———. "Flash: 99% Bad." *Alertbox.* 20 October 2000. Accessed 23 January 2009. http://www.useit.com/.

————. "How Users Read on the Web." *Alertbox.* 1 October 1997. Accessed 23 January 2009. http://www.useit.com/.

————. "Original Top Ten Mistakes in Web Design." *Alertbox.* May 1996. Accessed 23 January 2009. http://www.useit.com/.

Raymond, Eric S. "The HTML Hell Page." Accessed 15 March 2009. http://catb.org/esr/.

Venturi, Robert, Denise Scott Brown, and Steven Izenour. *Learning from Las Vegas: The Forgotten Symbolism of Architectural Form.* Rev. ed. Cambridge, Mass.: MIT Press, 1977.

Wagner, Carl. Personal interview. 2 November 2007.

Wusteman, Judith. "Formats for the Electronic Library." *Ariadne* 8 (1997). Accessed 15 March 2009. http://www.ariadne.ac.uk/.

8   \<frame\>ING REPRESENTATIONS OF THE WEB

MICHELLE GLAROS

Frames divide a page into a series of windows (or "cells"). Via \<frameset\> *tags, and with the use of* src *and* name *attributes, one page can embed one or more pages as if they constitute a single page. Individual windows in a frameset can be swapped using the name and target attributes. Although frames are no longer widely used, they actualized Douglas Engelbart's 1963 concept of a multiwindowed interface in which users could navigate various pages of information at once. Before Netscape version 2.0's release in 1995, Web users who did not upgrade their browsers often found the message "Your browser does not support frames" awaiting them on the screen. Today the most common examples of frames are navigational. Google's Image Search displays search data in the top frame, and the page that contains the found image below. Web mail clients divide the page to provide menu options, a message list, and/or an inbox. Content management systems often use frames to separate authoring elements from rendered code. In this manner, the function of juxtaposing different pages is replaced with the management of a single page in several forms.*

> \<frame\>:
> - a structure for admitting or enclosing something
> - usually, frames (used with a plural verb) the framework for a pair of eyeglasses
> - a particular state, as of the mind
> - one of the successive pictures on a strip of film
> - to form or make, as by fitting and uniting parts together; construct
> - to conceive or imagine, as an idea
> - Informal. to incriminate (an innocent person) through the use of false evidence, information, etc.
>
> DICTIONARY.COM

<frame>, A BASIC HTML TAG used to delineate and manipulate the writing spaces of the browser, has been denigrated by usability experts and effectively excommunicated from Web design. Supplanted by Cascading Style Sheets and content management systems, frames became archaic, a vestige of markup language that no longer holds currency in contemporary usage and design. The tag <frame> has disappeared from most current uses of markup language. The frame, however, has not. Literally and figuratively, Web writing, design, and representation remain framed in a number of ways that are both generative and restrictive. *Frame* is still a keyword for considering the creative and aesthetic development of media; our conceptualizations of media are framed in a number of ways. With this essay/entry, I seek to explore a number of these framings, illuminating the tensions that stretch between them and considering the opportunities afforded by certain reframings.

   <frame> allowed for the simultaneous display of multiple URLs onscreen but cut loose the anchors built into browsers. The metatag <frameset> governed these subordinate frames, carving the space of incription into discrete units while also contextualizing those units by determining their boundaries and metonymic partners. Ironically, designers usually deployed this provocative tag, which at times invited radical and unexpected juxtaposition, to tame the spaces of the browser. The sides of browser windows generally displayed stabilized menus that framed a changing, centered space. This popular and functional layout persists today despite the disappearance of the executable code that initiated it; see any number of news or shopping sites for examples of similar uses of Web space. In the mid-1990s, Jakob Nielsen, a prominent usability expert, declared that "frames suck" because they rendered bookmarks useless and produced nonpersistent URLs. <frame> inadvertently unleashed the browser and the viewer in such a way that we sometimes became lost, unable to locate ourselves and our perceptions in the ways to which we had become accustomed. Such inadvertent sabotage pointed to the liberating potential framed by this tag as well as the core tension that separates electronic writing from print: the tension between taming and unleashing the space of writing. By severing the ties that bind specific content to browser space and rendering URLs nonpersistent, frames cast users adrift on a hypertextual sea, locating and dislocating at once. The disturbance caused by <frame> illuminates the paradox of our desires. When we want to find information, we prefer straightforward, locatable content that can

easily be referenced and rereferenced. Yet we are simultaneously disappointed with a Web that mimics the phone book, road atlas, newspaper, and encyclopedia. We feel cheated and oversold because we also desire the Web that much new media theory heralds—one that is a creative, inventive studio where we encounter unexpected forms and expressions. Both of these desires are in fact framings of the Web, representations of the Web's promise and potential.

This essay/entry seeks to recuperate the tag <frame> by reconsidering its limitations and promise in the context of a number of possible permutations. The tag carries and displays a host of meanings, and like ideology that also renders itself invisible to its subjects, our perception and understanding of electronic writing remains framed in a number of often unacknowledged and unrecognized ways. Calling us to consciousness, <frame> directly challenges us to bear in mind what it means to frame and to be framed.

</writing>

My exploration of the frame and its influence on electronic writing began with a personal experience that turned my attention away from the language arts and toward the plastic arts. When I took a job at a small public university that was at one time a normal school but had become a school whose mission focuses on digital technologies, I organized a series of electronic writing workshops as a faculty development opportunity. In organizing these workshops, my goal was to create an environment in which language arts faculty could share not only their pedagogical uses of computing in the classroom, but also their theoretical research and thinking about the machines in our midst. At our first meeting, I introduced the group to George P. Landow's work with the Victorian Web at Brown University. I purposefully selected this work as our focus because it is well documented in the professional literature of the language arts, because it is conceptually accessible yet provocative (that is, everyone attending the workshop would have some familiarity with Victorian studies, albeit not with the Web's reframing of literary studies), and because it was written by students. Working at an institution that primarily concerned itself with teaching, I thought the latter consideration to be important because it provided a necessary hook that more esoteric instances of electronic writing do not always make readily apparent. Nevertheless, I was simultaneously concerned that the Victorian Web was too pedestrian, not provocative enough to really inspire my colleagues. To my surprise, soon after we began exploring the Victorian Web, one of the senior writing professors turned to me and asked quite pointedly: "Where's the writing?" I must admit this question perplexed me: did she mean that the sample hypertext was too ordinary to qualify as quality writing, or did she mean that the Victorian Web was too unusual to be writing?

As I considered how my colleague might be framing her question, I was reminded of a similar scene. Upon first confronting Marcel Duchamp's readymades in the gallery, art patrons retorted, "Where's the art?" Duchamp's readymades called on patrons to reflexively consider the nature of the art object, suggesting that everyday industrial objects might indeed hold artistic value and that the romantic aesthetic of art, artists, and artistic production might be an ideological frame worth interrogating. Might a mass-produced urinal be art? The readymades also interrogated the supposedly neutral space of the gallery, showing that space to be a frame that projected meaning, value, and status onto the objects contained within. A mass-produced urinal might indeed be deemed art if it is installed on a pedestal and housed in a gallery. My colleague's explanation of her question aligned her with Duchamp's early critics, and it proved to be quite telling because it indicated that she and I see very different things when we look at the same screen. She sees a list, an image, a few sentences or phrases, but no writing. For her and readers like her, lengths of alphabetic text define writing. I see that there are other ways to frame or define writing: in every design choice—in each choice of color, font size, and style; in each choice of word and image; in the layout, ordering, and framing; in the decisions to include and exclude; in the decisions to leave the blank spaces blank—in short, in the construction of every pixel and the manipulation of every space. The history of writing, as documented and theorized by scholars such as Walter Ong and Elizabeth Eisenstein, shows us that writing has been framed in a number of ways and that these frames have been both technological and ideological. The tablet, scroll, codex, and computer have all framed the concept of writing differently, as have the treatise, essay, novel, and hypertext. As philosophers from Jacques Derrida to Stanley Cavell, media critics from Bill Nichols to John Fiske, and art historians from Ernst Gombrich to John Berger continually remind us, all conceptions and perceptions, all frames are ideological. The questions, "where's the art? where's the writing?" direct us to once again reframe the concept of writing for colleagues and students who have come to imagine writing as nothing more than something that is defined by large chunks of alphabetic text inscribed on paper or screen space and as a skill to be mastered rather than an art to be practiced.

In its infancy, electronic writing promised to break these frames; it was hailed as an innovative technique that promoted a multilinear aesthetics combining text, image, and sound in ways never before possible. Now in its adolescence, electronic writing disappoints critics such as Ollivier Dyens because it appears to have failed to live up to its storied potential (in and of itself a frame composed by poststructural theory). In his "Artistic Statement" to the empyre Listserv, Dyens cites the domineering influence of the page, both a rhetorical and aesthetic frame, as the source of this failure:

> As much as I believe in the potential of the web, as much as I can imagine it bringing us new philosophical and artistic understandings of our culture, I am often disappointed by what this Promised Land offers. . . . As much as web sites are different from anything else we've seen or experienced before, we (most of us) still imagine them as big, electronic books on which to write, from which to read, books with pages and bookmarks and authors and editor-in-chiefs. . . . Even the terminology is wrong (web pages, electronic magazines, on-line newspaper).

With his critique, Dyens calls us to consciousness of the print-related concepts that frame much electronic writing: "Web pages, electronic magazines, on-line newspaper." In her essay "Electronic Literature: What Is It?" N. Katherine Hayles further explicates the influence of print; she notes that electronic writing includes a diverse set of practices including hypertext fiction, network fiction, interactive fiction, and location narratives as well as installation pieces, code work, generative art, and the Flash poem. The dominance of narrative forms in this list illuminates another frame that haunts electronic writing and disturbs Dyens; maintaining "such conventional narrative devices as rising tension, conflict, and denouement in interactive forms where the user determines sequence continues to pose formidable problems for writers of electronic literature" (Hayles). The Electronic Literature Organization's definition of electronic writing as "works with important literary aspects that take advantage of the capabilities and contexts provided by the stand-alone or networked computer" links the practice and the work to literary criticism, an intellectual practice that critics such as George P. Landow, Jay David Bolter, and Gregory L. Ulmer demonstrate is thoroughly interpolated by the philosophies of print. Even a concept as simple as the Aristotelian plot structure of beginning, middle, and end constitutes a frame that identifies and legitimates electronic writing as literature while simultaneously, and perhaps unnoticeably, constraining such work.

<cinema>

Likening electronic writing to early cinema in which filmmakers framed filmmaking with the language of theater, producing films that were simply recordings of theatrical performances, Dyens argues that we have yet to invent a language for electronic writing.

> We, in fact, replicate what was done in the early part of the 20th century when cinema was just recorded theater (the camera would not move, nor turn and the actor would come in and out of the frame, just as they would do on a stage). Most literary experiments have not worked (and still do not work) because most of them have only borrowed their language. We are still looking for our Eisenstein, for someone to clearly articulate what the web is, what its language is, what it evolves into, to clearly enunciate how it dissolves and then remodels our understanding of the world.
>
> What could be such a language?

The emergence of classical Hollywood cinema, like Duchamp's readymades, called artists and viewers to question the role and effect of the frame, reframing film as narrative cinema rather than as theater. In the 1920s, Soviet Montage again challenged the framing of film as narrative cinema with the development of intellectual montage, what Dziga Vertov called a pure filmic language, one framed not as theater or a language art but as a filmic art.

In early cinema, before the development of the principles of continuity and other cinematic rhetorics, the camera never moved into the performance space. The lens maintained one focal length throughout the recording; that is, the frame of the camera simply replicated the frame of the proscenium stage, recording the image as if the camera were a patron sitting in the middle of the theater. Film studies credits D. W. Griffith and his breakthrough films *Birth of a Nation* and *Intolerance* with the initial invention of a filmic language. In these films, Griffith moves the camera into and about the set, at times using close-ups that direct the viewer's attention to visual details. In addition, Griffith invented a new form of editing in which he cross-cuts between two simultaneous and ongoing events, effectively articulating the first chase scene and creating suspense through the manipulation of narrative space and time. Although Griffith is credited with producing the most fully conscious and accomplished early films that use this cinematic language, other early filmmakers were also beginning to recognize the fundamental concept that the shot, not the scene, can be a basic unit of meaning in film. This language of film, developed as a syntax called classical Hollywood cinema, uses composition and the careful ordering of shots to establish clear continuity of time, space, and narrative logic. This framing of cinema works to carefully and continuously locate the viewer in the space and time of the story world.

The principles of continuity are not the only language variation known to film, however. Soon after the development of these principles, Soviet Montage emerged as an alternative approach to filmmaking. Whereas early American filmmakers developed the principles of continuity in the service of narrative progression, many Soviet filmmakers of the 1920s used disjunctive editing as a method of organizing the symbolic meaning of the entire film form. Soviet Montage elevates editing and the juxtaposition of thought-provoking combinations of images above and beyond the service of the narrative, often disrupting the continuity of time, space, and narrative logic and casting viewers adrift in the story world. With Sergei Eisenstein as its most prolific practitioner, Soviet Montage emerged as a film language born of dialectical theory. Eisenstein and his peers shifted aesthetic attention to the cut as the key unit of meaning, developing it into a syntax called intellectual montage.

Whereas classical Hollywood cinema framed the fundamental unit of its filmic language as the shot and Soviet Montage framed the fundamental unit of its filmic language as the cut, electronic writing, as Dyens suggests, has yet to frame its own language. Usability experts such as Jakob Nielsen frame Web writing as a stable page, advocating for the language of print or the codex, a language that allows users to locate themselves in the text. Such framing mimics the goals of continuity filmmaking and ensures that viewers never feel lost; they can always identify and locate the space and time of the story world. Literary hypertext theorists such as Mark Bernstein suggest that the link is the basic unit of meaning for hypertextual writing, and although the link may be more inventive than the page, the link replicates the cut and thereby restricts electronic writing to the language of cinema. Framing hypertext language as the link aligns electronic writing with Soviet Montage, much as early cinema framed film as a sequel to theater. As Dyens notes, we are still confronted with the challenge of conceiving hypertext through a different frame, one that will generate its own language and syntax.

Dyens explores a potentially generative reframing of electronic writing when he argues that Net art, rather than the Web writing or even literary hypertext, has most successfully "let go of the old structures of narrativity and linearity." Hayles presses this same consideration when she notes that "the demarcation between digital art and electronic literature is shifty at best, often more a matter of the critical traditions from which the works are discussed than anything intrinsic to the works themselves." Although Dyens argues that the influence of print is more palpable than Hayles suggests, both point to a difference produced by framing such cultural production as art rather than literature. Net art, a form of digital art that emerged from the intersection of visual arts with programming, descends from a different lineage, one that suggests an alternative way of framing electronic writing. "Net art websites," notes Dyens,

> might not have found a truly specific net language yet but they have let go of the old structures of narrativity and linearity. Interesting net art websites treat the text as something to be perceived instead of read, as something to be "felt" and experienced and not as something to focus on. Net art websites truly understand that the web is an environment, something that has volume (intellectual, informational and emotional) and that one must build one's thought and reflection within this volume.

This analysis frames Net art as a particular inflection of the plastic arts— installation art: "something to be perceived . . . to be 'felt' and experienced . . . something that has volume."

What might it mean, then, to put aside the traditional frame of the language arts and take up instead the frame of the plastic arts when exploring electronic writing? If, as Bolter has persuasively suggested, writing with spaces distinguishes electronic writing from print, might we not be served by invoking installation art, and its reframing of the very concept of art, as an ancestor of electronic writing? Perhaps this art of space, a descendent of the plastic arts that takes space and place as both material and medium, might offer language artists inspiration for reframing writing without the page. Joan Campas begins to explore this reframing in her essay "The Frontiers between Digital Literature and Net.art" by putting the place of writing into play while also calling for the dematerialization of electronic writing. Her conception of electronic writing, informed by Roger Chartier's work, suggests "that the electronic representation of texts entails new relationships with writing, where the materiality of the book is replaced by the immateriality of texts 'with no place of their own.'"

In her book, *Space, Site, Intervention: Situating Installation Art,* Erika Suderburg provides a provocative introduction to the conceptual, philosophical, and ideological bases of installation art; her introduction informs this consideration of the frame of electronic writing in provocative ways. During the 1960s, the distinction between artwork and art criticism collapsed, and art was reframed as concept rather than object. Installation emerged as an unstable art form, provoking art historians and archivists to again question, "Where's the art?" as each new act of installing a work fundamentally changed the work itself. Installation, as a preparatory, pre-show practice, referred to the act of hanging or arranging works of art in the supposedly neutral spaces of galleries or museums. As an artistic practice, however, installation redeploys this mundane practice as an act of creative production. Like Duchamp's readymades, installation art interrogates the frame, making visible what initially appears to be neutral and even invisible. Today, as a developed art form, installation is an art that consciously manipulates, configures, or inscribes—that is, reframes—spaces (both public and private), showing them not to be as neutral as we might think. Installation art is a site and temporally specific practice. Indeed, with installation, space becomes material rather than medium. Installation artists sculpt spaces using a variety of tools including light and audio, performance, architecture, narrative, and video. Installation art often invites its viewers or readers to enter into the space of the art and become a part of the piece. Through the manipulation of space, such art inscribes the viewer/reader's aural, spatial, visual, and environmental planes of perception and interpretation. The work is informed by activities that reconfigure the body's relationship to the space it occupies and consequently reformulates and continually rewrites. Suderburg notes that Robert Smithson's use of the terms *site* and *nonsite* to label his works that removed samples from exterior sites and placed them into the "neutral" space of the gallery demanded an expansion of what could be thought of as art. Content could be space, space could be content, as sculpture was extrapolated into and upon its site. As each work is reinstalled in a new space, it reframes and is reframed by that space; each subsequent act of installation calls attention to the fact that the old work is becoming a new work. In addition, installation art reframes the observer as material, as participant; installation reframes art space as social space and social space as art space. The participant-observer (or reader) becomes implicated in the work in a manner that differs considerably from the conventional relationship between the viewer of painting or sculpture whose frame (or pedestal) separates the art object from the patron.

Net art, as Dyens suggests, may be more successfully inventing a new digital language (such as the filmic languages of classical Hollywood cinema and Soviet Montage) because the critical tradition that frames such work already reflects a shift from object to concept. The question, "Where's the writing?" suggests that composing with image, sound, and text in digital media might mean setting aside traditional distinctions between the language and the visual arts. For language artists, it might mean considering space as seriously as we have traditionally considered time. G. E. Lessing's extended essay, *Laocoon: An Essay on the Limits of Poetry and Painting*, teases out this distinction. Lessing argues that while space is the domain of the visual arts (exemplified by sculpture and painting), time is the domain of the ever-dominant literary arts (exemplified by poetry). The visual and the literary arts, sister arts, can be compared but never reduced into one another; their respective frames, space and time, mark their fundamental difference. During the last decade of the twentieth century, theorists and critics of new media continually suggested that one of the important changes wrought by computing was the fierce hybridization of the literary arts. No longer was literature primarily the domain of the word; with new media, literature became the domain of word, image, and sound. In current practices of Net art, we see this prophecy fulfilled. New media theorists from Jay David Bolter (*Writing Space*) to Lev Manovich (*The Language of New Media*) suggest that the key change computing brings to literature is the foregrounding of space over time, thereby challenging Lessing's postulate that when properly practiced, the literary arts aesthetically manipulate time over space.

</frame>

Installation and electronic writing share certain affinities. Both sculpt spaces with tools such as light, audio, image, and process. Both invite the viewer/reader or user to enter the space in which the work is being composed. Yet our imaginations are trapped by traditional conceptions of the language arts, which prize clarity and linearity. We are like those early filmmakers who found it difficult to escape the language of the stage. My goal is to reframe the ideological expectations I bring to bear when I contemplate writing in, with, and through digital media. If the notion of "writing" constrains my imagination, unconsciously tying me to the language of print, my gamble is that I might shatter that frame if I imagine sculpting (rather than composing) in, with, and through digital media as an act of installing rather than writing. Perhaps then we will invent new languages for new media and find our D. W. Griffiths and Sergei Eisensteins.

Such arguments mark an important reframing of writing implying that electronic writing is more closely related to contemporary art practices exploring space in time. In their collaborative project *Remediation: Understanding New Media,* Jay David Bolter and Richard Grusin move away from the modernist tendency to define essential characteristics of new media and instead argue that all media work by remediating, by translating, re-fashioning, and reforming other media. Following Bolter and Grusin's lead, might we resurrect and remediate the HTML tag <frame> as an executable code that continuously calls us to consciousness of the lens through which we approach electronic writing and the representations of the Web? Such a tag could invite us to frame installation art, a practice of conceptual art, and its peculiar means of engaging space as medium and space in time as a remediated ancestor of both new media art and electronic literature.

Recent developments in electronic writing suggest that such a reframing is both appropriate and necessary for continued critical development and creative exploration. Writing.3D, a form of three-dimensional electronic writing that operates on the z-axis, at times simulates the perception of depth or space (through techniques familiar to painting and cinema, including linear perspective and overlapping edges), but in some instances, such as CAVE projects, it literalizes the idea of electronic writing as installation (Raley). CAVE projects are site-specific instantiations of electronic writing that take advantage of Brown University's CAVE, a "three dimensional space in which the user wears virtual reality goggles and manipulates a wand" (Hayles). A number of contemporary electronic literary artists, including Robert Coover, Noah Wardrip-Fruin, and Talan Memmott, have created CAVE works that "enact literature not as a durably imprinted page but as a full-body experience that includes haptic, kinetic, proprioceptive and dimensional perceptions" (Hayles). Mobile computing also challenges us to consider the possibilities and effects afforded by the tradition of unfixing the art/literary object. The unwiring of the Net demands another radical reconsideration of the place and space of writing. Portable screenic devices both tether such writing to the two-dimensional world of desktop computing (through the two-dimensional screen) while simultaneously liberating it from the desk/office/home/lap or CAVE, making the place of writing potentially the world. Not since the development of the codex have the transformative possibilities of mobilizing writing loomed so large.

The key features of Web 2.0 also echo the effects of installation art and the promise of hypertext. Web 2.0, with its tags, folksonomies, and RSS (Really Simple Syndication), allows users to write with data (and to write data sets), much as installation art allows viewers to participate and become part of the work. With Web 2.0 technologies, readers become writers not by choosing from a set of predetermined paths, but by associating files according to their own creative logic, with their own language of keywords, thereby creating a syntax that juxtaposes and recontextualizes those files through the repetition of such tags. Web 2.0 tags invite users to imagine and create their own associations, unlike HTML tags, which function as a fixed set of instructions from the Webmaster to the browser. Web 2.0 tags allow users to reconfigure the spaces that frame digital files as well as the spaces framed by digital files.

Digital technologies continue to challenge traditional conceptions of writing by putting them in relief. For fifty years we have witnessed writing from another vantage point, glimpsing the edges of the frame. We have been able to see its borders and boundaries, its curves and contours. The development of electronic writing has shone another kind of light on composition. And like the high-key light of film noir, this light recasts the language arts, the art of writing, in provocative ways. Electronic writing invites us to reconsider what writing is, how it is practiced, and how it is taught, calling us to remain conscious of the <frame>. Writing, an act of inscription, is neither medium nor form specific; it is an activity that frames the great play of language and place, art and space.

My ambition with this entry is to reengage writing with experiences outside the symbols, mythologies, and constraints of the conventional language of writing or composing, to free it from the shackles of its customary frame. This entry thereby proposes that we remember the tag <frame> as an executable code that reminds us to always contemplate and consider the effects of framing. This entry also suggests that we use installation art as a model for electronic writing or new media art compositions. A 4-D art form that arranges and inscribes (and thus "writes") space and time, installation art uses sound and image to sculpt particular "scenes." In breaking out of the frame and sliding off the pedestal, this art of space offers language artists inspiration for imagining writing without the page. Electronic writing likewise reframes and reconfigures our approach to texts while redefining the community of readers/viewers able to share its meanings and values. By carving up and reconditioning the information field, such creative practices disrupt traditional ways of understanding space, time, and space in time; they both reflect and provoke changes in how we think and write.

Works Cited

Berger, John. *Ways of Seeing*. London: Penguin, 1972.

Bernstein, Mark. "Patterns of Hypertext." In *Proceedings of Hypertext '98*, edited by Frank Shipman, Elli Mylonas, and Kaj Groenback, 21–29. New York: ACM, 1999.

The Birth of a Nation. DVD. Directed by D. W. Griffith. 1915. Los Angeles, Calif.: Kino Video, 2002.

Bolter, Jay David. *Writing Space: Computers, Hypertext, and the Remediation of Print*. Hillsdale, N.J.: Lawrence Erlbaum Associates, 2001.

Bolter, Jay David, and Richard Grusin. *Remediation: Understanding New Media*. Cambridge, Mass.: MIT Press, 2000.

Campas, Joan. "The Frontiers between Digital Literature and Net.art." *Dichtung-Digital*. 24 February 2004. Accessed 7 December 2008. http://www.brown.edu/.

Cavell, Stanley. *Must We Mean What We Say: A Book of Essays*. New York: Scribner, 1969.

Chartier, Roger. *Historia de la Lectura en el Mundo Occidental*. Madrid: Taurus, 2001.

Derrida, Jacques. *Of Grammatology*. Translated by Gayatri Chakravorty Spivak. Baltimore: Johns Hopkins University Press, 1977.

Dyens, Ollivier. "Artistic Statement." *Empyre Listserv*. 16 January 2002. Accessed 15 March 2007. https://mail.cofa.unsw.edu.au/.

Eisenstein, Elizabeth. *The Printing Press as an Agent of Change: Communications and Cultural Transformations in Early Modern Europe*. New York: Cambridge University Press, 1979.

Electronic Literature Organization. Accessed 5 December 2008. http://eliterature.org/.

Fiske, John. *Power Plays Power Works*. New York: Verso Books, 1993.

Gombrich, Ernst. *The Story of Art*. London: Phaedon, 1950.

Hayles, N. Katherine. "Electronic Literature: What Is It?" *Electronic Literature Organization*. Version 1.0. 2 January 2007. Accessed 12 December 2008. http://eliterature.org/.

Intolerance. DVD. Directed by D. W. Griffith. 1916. Los Angeles, Calif.: Public Domain Flicks, 2008.

Landow, George P. *Hypertext 2.0: The Convergence of Contemporary Critical Theory and Technology*. Baltimore: Johns Hopkins University Press, 1997.

———. *The Victorian Web: Literature, Culture, History in the Age of Victoria*. June 1995. Accessed 10 June 2005. http://www.victorianweb.org/.

Lessing, G. E. *Laocoon: An Essay on the Limits of Poetry and Painting*. Translated by Edward Allen McCormick. Baltimore: Johns Hopkins University Press, 1984.

Manovich, Lev. *The Language of New Media*. Cambridge, Mass.: MIT Press, 2001.

Nichols, Bill. *Ideology and the Image: Social Representation in the Cinema and Other Media*. Bloomington: Indiana University Press, 1981.

Nielsen, Jakob. "Why Frames Suck (Most of the Time)." December 1996. Accessed 20 February 2007. http://www.useit.com/.

Ong, Walter. *Orality and Literacy: The Technologizing of the Word*. New York: Methuen, 1982.

Raley, Rita. "Editor's Introduction: Writing.3D." *Iowa Review*. September 2006. Accessed 10 January 2009. http://www.uiowa.edu/.

Suderburg, Erika. *Space, Site, Intervention: Situating Installation Art*. Minneapolis: University of Minnesota Press, 2000.

Ulmer, Gregory L. *Internet Invention: From Literacy to Electracy*. New York: Longman, 2002.

Vertov, Dziga. *The Kino-Eye: Writings of Dziga Vertov*. Berkeley: University of California Press: 1985.

9 BREAKING ALL THE RULES

`<hr>` *and the Aesthetics of Online Space*

MATTHEW K. GOLD

Largely forgotten by Web authors and readers today, the horizontal rule, `<hr>`, creates a horizontal line across a page. `<hr>` duplicates the function of rules in print, which create boundaries between blocks of text. Eventually, `<hr>` was duplicated in function by more ornamental rules, images added to a page in place of the horizontal rule. These images may have blinked, resembled ornate script, reproduced flowers arranged as borders, or displayed animals and other animated figures. Because users of early free hosting sites such as Tripod and Geocities often created single pages, rather than multipage sites, the ornate horizontal breaks were meant to divide content over a large vertical space. Contemporary Web practices have since made such rules unfashionable; Web sites that offer galleries of flashy, animated rules are a curiosity today. Overall, the questions raised by `<hr>` revolve around boundaries of all kinds: print/Web, image/text, form/content, semantic/presentational, amateur/ professional. `<hr>` demonstrates the writerly need to pose a boundary at some place on a given page, whether to divide content or to draw attention to the different media used within one specific Web space.

> That is why, beginning to write without the line, one begins also to reread past writing according to a different organization of space.
>
> **JACQUES DERRIDA,** *OF GRAMMATOLOGY*

IN 1997, DAVID SIEGEL DECLARED WAR on the horizontal rule (`<hr>`), a popular HTML tag used to create horizontal lines across Web pages. Siegel, a self-styled provocateur and graphic designer, had published successive editions of his popular Web-design guide, *Creating Killer Websites*, in 1996 and 1997. In that book, he heralded the rise of "third-generation site design," which purported to do away with the text-, icon-, and menu-driven layouts of the early to mid-1990s. The `<hr>` tag was among the first design elements he wished to discard; Siegel argued that the creators of early Web pages had overused horizontal rules to such an extent that the Web had become "littered" with "visual junk" (69). What had initially been pristine virtual space, he suggested, had devolved into a trash heap of superfluous pixels. Siegel called on future designers to "Banish Horizontal Rules!" (69); nothing less than a complete eradication of the tag would do.

Why did early Web designers use horizontal rules so frequently? And, given the near omnipresence of horizontal rules on the early Web, why have those rules received such scant attention in the years since? The general absence of commentary on the horizontal rule reflects a more general void in Web design history (a field of scholarship that is, admittedly, in its infancy). First- and second-generation Web sites tend to be viewed as aesthetic embarrassments, as evidence of how poor sites looked before graphic designers wrested control of the Web from programmers and technicians. As David Siegel noted, for instance, most early Web sites had all the allure of "slide presentations shown on a cement wall" (27). And the horizontal rule, the most banal design element of the most basic early Web pages, has come to embody all that is passé about early Web design.

The urge to bury the primitive designs of the early Web, or to ascribe their now-unfashionable aesthetics to the scientific and technical backgrounds of early Web site creators, has fostered narratives that sometimes contain a teleological trajectory, a kind of graphical determinism. In his seminal text *Writing Space*, for instance, Jay David Bolter charts a move from the original version of HTML, which gave designers little control over the layout of their pages, toward more recent efforts by graphic designers to manipulate not just the vertical, but also the horizontal visual axis of the page. "As a result," Bolter writes, "some of the most compelling Web pages began to look like magazine advertisements, with striking visual metaphors, display fonts, drop-shadowed texts" and other adornments (69). Although Bolter situates that narrative within a larger story about remediation and a smaller one about the professionalization of graphic design in the context of late capitalism, the "breakout of the visual" (47) that he describes positions early Web design as the rude ground from which more mature Web sites flowered.

Returning to the aesthetic scene of the early Web, however, can help us rethink the development of the medium itself, particularly in relationship to evolving notions of space in virtual environments. Early Web designers used horizontal rules to break up dense blocks of hypertext, thus dividing the browser window into zones of legibility and visibility, rationality and readability. What were the effects of this carving up of space, this segmentation of visual territory, this projection of rationalized grids upon a new medium where the nature of virtual space promised an escape from the physical logic of print? To what extent have the aesthetics of linearity, the right angles of squares and rectangles, the taxonomic imperatives of division and separation, and the "rules" of Western logical structures codified our ongoing sense of what the Web is, what it should look like, or what it could be? And to what extent have newer trends in Web-design aesthetics obscured the ways in which more modern Web sites continue to nourish linear structures of thought, linear patterns of social behavior, and linear modes of discourse—all in contravention of the supposedly decentered nature of the medium itself? Why, to appropriate a question Bolter asks in *Writing Space* (107), have we not yet reached the end of the line?

It is precisely because horizontal rules no longer seem relevant to contemporary discussions of Web design that this moment provides a fruitful opportunity to reconsider the tag. Through an exploration of the historical development of the horizontal rule, an examination of its antecedents in earlier forms of written and printed discourse, and a meditation on controversies surrounding the use of the tag in digitized space, this essay will situate the horizontal rule within the shifting aesthetics of lexical media. I do not intend to make aesthetic judgments about the tag itself; whether or not horizontal rules beautify the Web pages they adorn concerns me less than the place of the tag in our shifting understandings of the aesthetics of virtual space.

Setting the Rules

The computer scientists who created the HTML markup language conceived of it as a structural, rather than a presentational, language. Its main purpose, for Web architects such as Tim Berners-Lee, was to enable scientists to share their research with one another (Bolter 69). The ability of HTML code to adapt to the user's particular display settings or browser version was welcome evidence of its versatility. But such visual instability frustrated graphic designers such as Siegel, who deplored the lack of control that HTML gave him over the visual appearance of the Web sites he created. Allowing visitors to determine the display of a Web page, Siegel fumed, was "like telling the artist how to hold the brush!" (4). He reported that "a wave of panic swept through" him the first time he saw one of his Web sites displayed differently across two browsers (4).

Tim Berners-Lee and Daniel Connolly added the horizontal rule to the HTML language in the Network Working Group's 1995 description of HTML 2.0, the first official standard of the Hypertext Markup Language. The two authors gave the tag a prominent placement in their list of HTML elements, in between the tags for line break (`<br>`) and image (`<img>`). The description of the tag itself, however, was relatively nondescript: "The `<HR>` element is a divider between sections of text; typically a full width horizontal rule or equivalent graphic" (Berners-Lee and Connolly). Berners-Lee and Connolly defined the element through its functionality as a border between "sections of text." Though the horizontal rule appeared as a visual element on the hypertext page, the authors described its look self-referentially: a horizontal rule looked . . . like a horizontal rule—except, perhaps, when it resembled an "equivalent graphic." While this lack of descriptive precision perhaps opened up a welcome range of possible visualizations of the tag, it also buttressed Siegel's contention that the "Framers of the Web" were less interested in design than in functionality.

A decade and a half later, the most recent iteration of HTML 5 defines the horizontal rule with surprising exactness:

> The hr element represents a paragraph-level thematic break, e.g. a scene change in a story, or a transition to another topic within a section of a reference book. (Hickson)

This description of the tag continues the earlier emphasis on functionality; `<hr>` is conceived of as a break or a rupture in a document. But the specification is oddly specific; it places the horizontal rule in the context of paragraphs, stories, and reference books, as if horizontal rules do not appear, or do not appear to have meaning, outside of such narrative frameworks.

Horizontal rules lack many elements of what we have come to understand as fundamental aspects of hypertext, which George Landow has described as "texts composed of blocks of text—what Barthes terms a *lexia*— and the electronic links that join them" (3). The horizontal rule would seem to fit more properly under the term *hypermedia,* defined by Landow as an extension of hypertext that includes "visual information, sound, animation, and other forms of data" (3). And yet early horizontal rules lacked many of the functions that we associate with hypertext and hypermedia: unlike links, horizontal rules could not be clicked; unlike images, they could not be downloaded; unlike tables, they could not contain information or other page elements. In an interactive medium, horizontal rules refused all forms of exchange.

The horizontal rule itself fits strangely between the realms of text and image, coming close to what W. J. T. Mitchell has termed the *imagetext*.[1] According to Mitchell, imagetexts are "composite, synthetic works (or concepts) that combine image and text" (89n). Although horizontal rules appear on Web pages as images, they are not downloaded by the browser, but rather are created through its GUI, making them unlike most binary images on the Web. The horizontal rule is also text, created in the source code of the page through the typing of four textual characters: `<hr>`.

Despite the blended image/text/code properties of horizontal rules, Berners-Lee and Connolly's tautological definition of the tag implied that the horizontal rule was known, obvious, and self-evident—that it needed no elaboration or explanation. If this was so, it was because rules had long been used by writers and printers to divide, order, and rationalize space on the printed page.

Antecedents of the Horizontal Rule: A Selective History

Dividing Space in Early Writing Systems

In some of the earliest forms of writing, such as the Paleolithic paintings in the caves of Lascaux, images of horses, bison, and stags converged on large visual fields that were undivided by lines. But as the Sumerians living in Mesopotamia began to develop their pictographic writing systems, they increasingly used lines and grids to order symbols on their compositional surfaces.

The pictographic clay tablets created in the Mesopotamian city of Uruk (Figure 9.1), for example, used horizontal and vertical rules to separate one set of symbols from another (Meggs and Purvis 6–7). On such tablets, lines carved into the stone divided the writing space into distinct zones of meaning. In Figure 9.1, for instance, three vertical lines at the top of the stone created four textual spaces of roughly uniform dimensions, each of which contained one to four pictographs; a long horizontal line divided the vertical space of the tablet in half, creating a larger lower zone. The increased size of the symbols in this lower zone, combined with the abundant blank space around them, announced the central importance of the lower zone to the overall meaning of the tablet.

The Sumerians used their tablets to keep track of increasingly complex social, religious, and economic relations that necessitated the storage and

Figure 9.1. Limestone tablet with Sumerian pictographic script, end of the fourth millennium, Mesopotamia. The horizontal and vertical lines divide and rationalize textual space into zones of meaning. Photograph by Erich Lessing/Art Resource, New York.

retrieval of information. Typically, they used clay tablets to record financial transactions, crop yields, and food storage inventories (Meggs and Purvis 6). As scribes began to record this information on wet clay tablets that were later baked into a hardened state, they used lines to organize their data.

In later hieroglyphic writing systems, horizontal lines separated visual space into distinct zones of meaning that reflected larger cultural belief systems. In her analysis of the Narmer Palette—a two-foot-long slate tablet created during the predynastic period of Egyptian history—Carol S. Lipson finds that Egyptians used register lines to mark the blank tablet. The carved lines on the tablet divided it into separate sections, each of which was ordered in a hierarchical manner: the top section represented the divine realm; the middle section contained images of the king; and the lowest section depicted the kingdom's vanquished enemies. Lipson finds that these three "windows" on the tablet created a syncretic, rather than sequential, narrative: each section of the tablet represented a distinct social space, but all were connected both visually and thematically to the larger theme of the king's relationship to the divine and mortal realms. That the lines creating these windows served to draw content together into a coherent narrative, to connect textual spaces even as they served to separate them, suggests that horizontal rules might have performed a similar function on modern computers, which are dominated by the windowed style of the graphical user interface (Bolter and Grusin 31).

In medieval illuminated manuscripts, scribes decorated their texts with complex illustrations or decorated letters, but they rarely used rules to separate blocks of text. The use of horizontal rules increased in the centuries after the invention of the printing press, although Gutenberg himself did not use them in his Bible (Arnold 46). They were used not just to separate and divide content, but also to decorate printed texts. This secondary function of decoration is particularly important to an understanding of the ways in which Web designers have used horizontal rules on their Web pages.

Horizontal Rules in Printed Books

Textual scholars consider printed rules to be typographical ornaments, graphical designs that adorn or embellish printed texts. In addition to rules, which are straight lines that might be ribbed, shaded, decorated, or swelled, the category of printed ornaments also includes headbands, tail-pieces, initials, border graphics, pointers, paragraph markers, and small illustrations (Ryder 13). In her *Dictionary of Colonial American Printers' Ornaments and Illustrations,* Elizabeth Reilly has noted that ornaments "were to movable type what decorated initials and borders were to script" (xv): they provided printers with a way to beautify their texts and to add unique design elements that printing presses made increasingly uniform.

Such ornaments appeared in books virtually from the invention of the printing press; for some time after the invention of the printing press, they were engraved on wood or metal and stamped into a book after the mechanical production of the text itself (Luidl and Huber 13). By the beginning of the seventeenth century, typefoundries began to cast ornaments in metal and to sell them to printers. In colonial America, where printers often relied on European typefoundries for their ornaments, the number of decorations that a printer had at his disposal became a register of his success as a printer (Reilly xviii). Along with other tools of the trade, printers gave ornaments to their descendents as heirlooms so that family businesses could continue to flourish (Reilly xx).

By the nineteenth century, plain, decorated, and swelled horizontal rules appeared regularly on both the title and inside pages of books. Plain horizontal rules were straight lines of varying thickness that might appear ribbed or shaded; often, two rules were printed closely together, with the top line having extra shading. Decorated horizontal rules could incorporate smaller graphics into a larger, line-shaped ornament; often, those graphics were based on natural forms, such as leaves, vines, or fruits, or on geometric forms, such as circles, diamonds, or squares (Figure 9.2). Swelled horizontal rules were lines whose enlarged centers tapered to sharpened points on both ends.

During the nineteenth century, horizontal rules appeared with great frequency on the title pages of American and European books, where they were set amid long titles that sometimes stretched to hundreds of words (McLean 7). Plain horizontal rules, for example, appear throughout the sample plates of Ruari McLean's *Victorian Book Design and Colour Printing,* a text that examines nineteenth-century printing and design in Britain. Although some publishers added ornate typographical objects to separate areas of content from one another, others used simple, horizontal rules to divide the visual space of the page and to segment meta-information about the book, such as titles and page numbers, from the main text itself.

Examples of both plain and decorated horizontal rules can be found in a wide variety of nineteenth-century texts, but two examples from mass-market texts will serve to illustrate the popular use of such ornamentation. Both plain and decorative horizontal rules appear in a late nineteenth-century edition of John Greenleaf Whittier's poems, published by the Worthington Company of New York in 1888. The edition was part of a series of texts written by popular authors; the books were moderately, though not cheaply, priced (Huttner). Both plain and decorative rules appear on the title page, which, unlike many texts of the period, contains not an

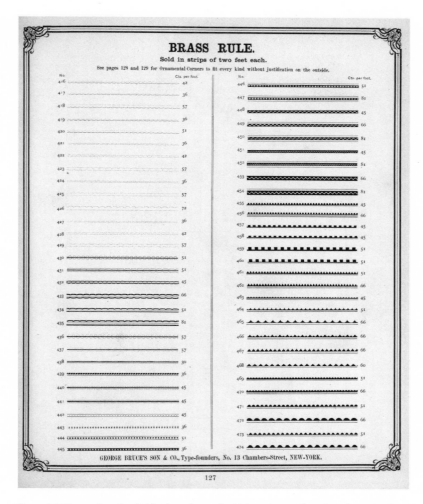

Figure 9.2. Brass rules. *An abridged specimen of printing types made at Bruce's New York typefoundry.* Illustration from George Bruce's Son & Co. (1869), 127. Courtesy General Research Division, The New York Public Library, Astor, Lenox and Tilden Foundations.

ornamental illustration, but rather a quotation (Figure 9.3). Short, plain horizontal rules also appear on the contents page and in the texts of the poems themselves (Figure 9.4). Plain horizontal rules also appear in a similar English edition of Oliver Goldsmith's poems and plays (Figure 9.5), where they separate the text of those plays and poems from bibliographic information such as the title of the book and the pagination.

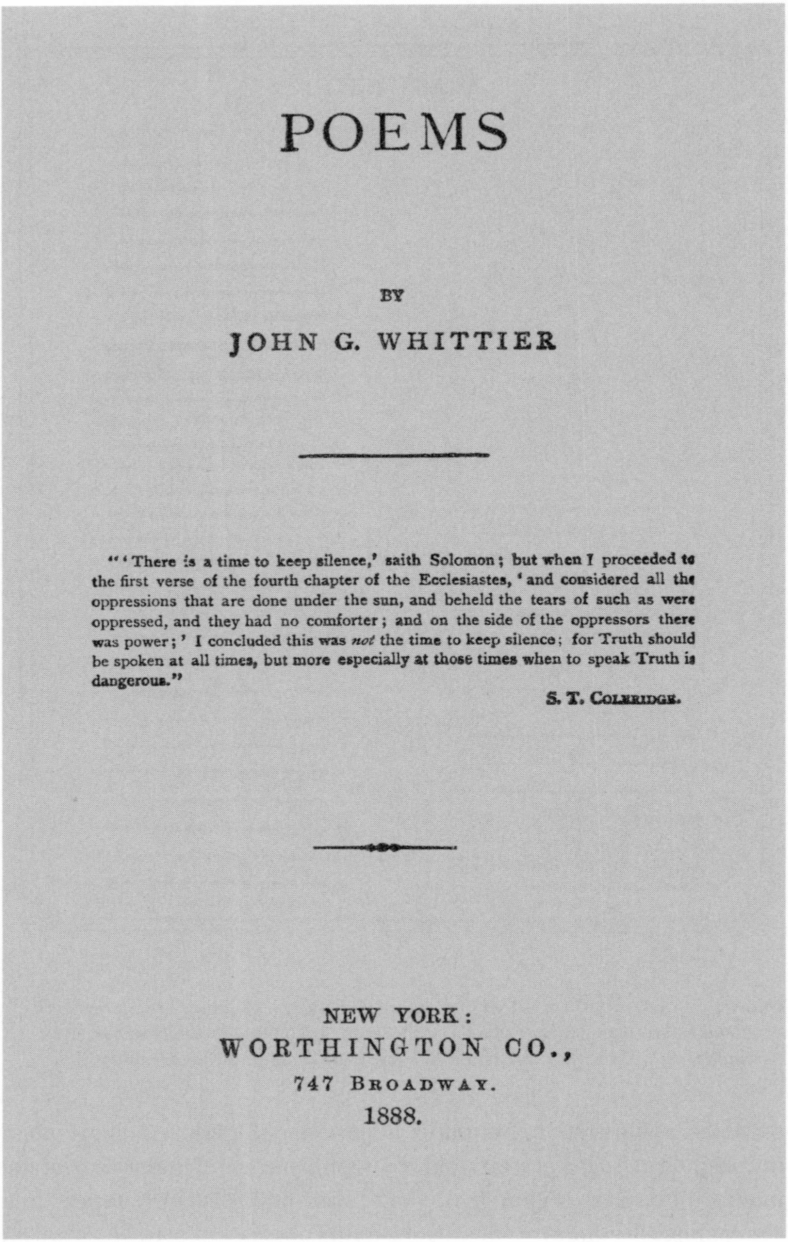

POEMS

BY

JOHN G. WHITTIER

"'There is a time to keep silence,' saith Solomon; but when I proceeded to the first verse of the fourth chapter of the Ecclesiastes, 'and considered all the oppressions that are done under the sun, and beheld the tears of such as were oppressed, and they had no comforter; and on the side of the oppressors there was power;' I concluded this was *not* the time to keep silence; for Truth should be spoken at all times, but more especially at those times when to speak Truth is dangerous."

S. T. COLERIDGE.

NEW YORK:

WORTHINGTON CO.,

747 BROADWAY.

1888.

Figure 9.3. Page containing one plain horizontal rule and one decorated rule. John G. Whittier, *Poems* (New York: Worthington Company, 1888).

TO GOV. M'DUFFIE.

" The patriarchal institution of slavery,"—"the corner-stone of our republican edifice."—*Gov. McDuffie.*

KING of Carolina—hail!
Last champion of Oppression's battle!
Lord of rice-tierce and cotton-bale!
Of sugar-box and human cattle!
Around thy temples, green and dark,
Thy own tobacco-wreath reposes;
Thyself, a brother Patriarch
Of Isaac, Abraham, and Moses!

Why not?—Their household rule is thine;
Like theirs, thy bondmen feel its rigor;
And thine, perchance, as concubine,
Some swarthy counterpart of Hagar.
Why not?—Like patriarchs of old,
The priesthood is thy chosen station ;
Like them thou payest thy rites to gold—
An Aaron's calf of Nullification.

Figure 9.4. Plain horizontal rules appear on an inside page of the text. John G. Whittier, *Poems* (New York: Worthington Company, 1888).

Mrs. Croak. And pray, what right then have you to my good-humour?

Croak. And so your good-humour advises me to part with my money? Why then, to tell your good-humour a piece of my mind, I'd sooner part with my wife. Here's Mr. Honeywood, see what he'll say to it. My dear Honeywood, look at this incendiary letter dropped at my door. It will freeze you with terror; and yet lovey here can read it—can read it, and laugh.

Mrs. Croak. Yes, and so will Mr. Honeywood.

Croak. If he does, I'll suffer to be hanged the next minute in the rogue's place, that's all.

Mrs. Croak. Speak, Mr. Honeywood; is there anything more foolish than my husband's fright upon this occasion?

Honeyw. It would not become me to decide, madam; but, doubtless, the greatness of his terrors now will but invite them to renew their villany another time.

Mrs. Croak. I told you he'd be of my opinion.

Croak. How, sir! do you maintain that I should lie down under such an injury, and show, neither by my tears nor complaints, that I have something of the spirit of a man in me?

Honeyw. Pardon me, sir. You ought to make the loudest complaints, if you desire redress. The surest way to have redress, is to be earnest in the pursuit of it.

Croak. Ay, whose opinion is he of now?

Mrs. Croak. But don't you think that laughing off our fears is the best way?

Honeyw. What is the best, madam, few can say; but I'll maintain it to be a very wise way.

Croak. But we're talking of the best. Surely the best way is to face the enemy in the field, and not wait till he plunders us in our very bed-chamber.

Honeyw. Why, sir, as to the best, that—that's a very wise way too.

Mrs. Croak. But can anything be more absurd, than to double our distresses by our apprehensions, and put it in the power of every low fellow that can scrawl ten words of wretched spelling to torment us?

Figure 9.5. Plain horizontal rules separate the main text from page numbers and title information. Oliver Goldsmith, *The Poems and Plays of Oliver Goldsmith, with the Addition of "The Vicar of Wakefield," "Memoir," Etc.* (London: Frederick Warne and Co., n.d.).

Newspapers began as pamphlets in the seventeenth century; the first weekly news periodicals were published in 1622. Initially, these news pamphlets had cover pages whose designs closely mirrored the title pages of books (Hutt 9). As we have seen in the previous section and in Figure 9.3, title pages during this period often incorporated horizontal rules, and early newspapers did so as well. As newspapers developed, textual space became increasingly cramped; late eighteenth-century editions of the *Boston Gazette,* for instance, divided content into three columns. Double rules were used at the top of the paper to highlight and set off, in relief, the date of publication (Hutt 26–27). Other early newspapers used rules to separate stories or advertisements, and they underscored headlines with decorative, swelled rules.

Typography has sometimes seemed to be a second thought in newspaper design; historian Kevin Barnhurst has characterized the newspaper as "inelegant and ephemeral . . . least among forms" (7). The crowding of content in newspapers is an important feature of that inelegance, but those design constraints emerged from cultural and political processes. In colonial America, for instance, the Stamp Act of 1765 effectively levied a tax against printers by requiring them to purchase official stamps for their publications. This put a premium on space and forced editors to cram as much text as they could onto the printed page (Hutt 47). In Europe, according to Allen Hutt, the front pages of early newspapers amounted to walls of solid type, an approach that he characterizes as "a solid, minimal-headed, set of editorial columns, its advertising back page presented [as] a sledge-hammer spread of big, bold display type" (48). Amid such large blocks of type, horizontal and vertical rules imposed some semblance of order, some means of organizing a brute mass of information.

In newspapers, horizontal rules have been used most frequently in the sense that Berners-Lee and Connolly described them: to separate and divide sections of text. As Edmund C. Arnold has observed, the easiest way to divide type masses is to draw a line between them. And yet he also notes that in newspaper design, rules were part of the functional apparatus of the press itself: in the era of hand-set newspapers, printers could use a rule not just as an optical divider, but also as a "physical fence to hold individual characters in line" (45). Rules quite literally kept the newspaper together.

Barnhurst has suggested that the tendency of newspaper editors to juxtapose various typographical styles—a design choice that partly accounts for the frequent use of horizontal and vertical rules on newspaper pages—is related to the rhetorical aims of the newspaper itself:

> A distinguishing trait of the newspaper during much of its history has been the intermingling of many typographic styles and schools. . . . The mixed typography parallels the jumble of content on newspaper pages . . . and may signal at least broadly an intention (or perhaps a pretension) to address a whole diversity of readers. (134)

Barnhurst's suggestion that the heterogeneous graphic design of the newspaper indicated the desires of its publishers to reach out to a "whole diversity of readers" also holds true for electronic writing and reading spaces. As Bolter and Grusin have observed, "many web sites are riots of diverse media forms" that "mix media and styles unabashedly" (9). And as Bolter argues, we have to value the ability of the Web to "promote multiplicity [and] heterogeneity" (204). Horizontal rules, which had helped separate and divide the writing space of the newspaper page into discrete sections, performed a similar function on Web sites.

The Web-Based Horizontal Rule

Having examined the print-based antecedents of the `<hr>` tag, we are now in a better position to examine the specific ways in which the horizontal rule remediated the typographical conventions of printed books and newspapers.

The Default Horizontal Rule in Early Web Design

In the mid-1990s, when slow 14.4- and 28.8-kbps modems limited the number and size of graphical elements that could be placed on Web pages, the horizontal rule allowed designers to create, with four keystrokes, an element that looked and functioned like an image of a line without requiring the downloading of any graphics. In an era before page layout was accomplished through tables or `<div>` tags, HTML pages typically consisted of a single vertical column. Web designers used horizontal rules to divide and separate blocks text, navigational elements, and images.

Siegel argued that the design of first-generation Web sites reflected the data-driven concerns of the scientists who pioneered the World Wide Web. Such sites were often structured in a linear fashion:

Looking at a typical first-generation page, you can see the restrictions imposed by slow modems, monochrome monitors, and the default browser style sheets. The page displays a top-to-bottom, left-to-right sequence of text and images, interspersed with carriage returns and other data-stream separators like bullets and horizontal rules. (12)

As Siegel notes, the horizontal rule fit squarely into the aesthetics of first-generation Web sites, where a combination of technical limitations necessitated rudimentary HTML page designs that reflected the rationalized logic of computer scientists. Text was perceived as data, and horizontal rules acted as separators of that data, elements that helped organize streams of characters into readable, regimented patterns of content.

In its early utilitarian function as a divider and separator of content, the Web-based horizontal rule remediated the rules of newspapers. The three-dimensional beveling of Web-based horizontal rules were elements of what Bolter and Grusin would call their hypermediacy. By encoding an illusion of depth into the Web page, horizontal rules called attention to

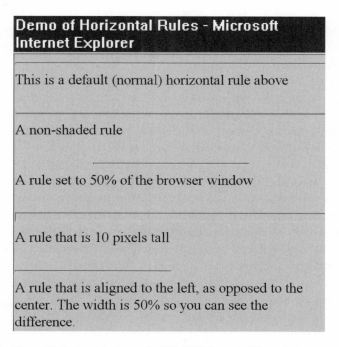

Figure 9.6. Horizontal rules on the Web, 1998. Image by Edmond Ho.

the ability of hypermedia to move beyond two-dimensional surfaces, to offer greater depth of field than printed texts. Even on the earliest Web pages, the beveled horizontal rule created seams in the page that presaged later, more complex and immersive three-dimensional Web-based graphics. Like another early, omnipresent Web-design trend, drop-shadowed images, the beveled edges of the default horizontal rule added a three-dimensional aspect to the perceptual experience of the World Wide Web.

The `<hr>` as Ornament: Extending the Rule through Attributes and Images

The horizontal rule tag could be extended through a number of attributes: its horizontal width, vertical thickness, color, alignment, and shading could be controlled in the HTML code itself. Although all of these attributes have been deprecated, or rendered obsolete, in HTML 4.0.1 (Jacobs), they allowed early Web designers to use the horizontal rule tag to create abstract geometric shapes that bore little resemblance to default horizontal rules. The horizontal rule tag, for example, could be extended to create a solid red square through the following tag: `<hr color="red" size="100" width="100" noshade>` (Figure 9.7). That users could extend the `<hr>` tag well beyond the simple geometric shape of a line suggests that the tag had the potential not just to separate and divide blocks of text, but also to decorate pages.

Indeed, Web page designers in the mid-1990s began to replace the horizontal tag with images that changed the nature of the horizontal rule itself. These often animated images converted the line of the horizontal rule into a moving element: in Figure 9.8, for instance, we can see a fire engine racing along a line to spray water on a raging fire, figures dancing across a collection of national flags, and a bicycle moving along a road. Clip-art Web sites collected such images in galleries that bore resemblances to the type-specimen books of the nineteenth century (compare, for instance, Figure 9.8 to Figure 9.2). Although these images were often amateurish in style—they appeared on Web sites with page titles such as "Hot HRs" and "Bells and Whistles"—they allowed for a new level of hypermediacy in the horizontal rule. These `<hr>`-like animations did more than place a beveled line on a page; they made the horizontal rule itself an activated element of the page layout and served not just to separate masses of text, but also to adorn them. One creator of an animated `<hr>`-like graphic proudly advertised his "Horizontal Rule with a Spinning Ball in the Center!" by urging visitors to "replace the `<hr>`'s in your page with

Figure 9.7. Horizontal rule configured as a solid square in pre-4.01 versions of HTML.

this bauble for a little eye-grabbing spice" (Woodman). Such <hr>-like animated images remediated not the plain horizontal rule of the newspaper, but rather the decorative typographical ornament of the printed book.

Condemnations of Web-based horizontal rules as garish and unappealing are best contextualized within larger debates about the role of ornament and of ornamentation more generally. In his study of architectural ornament, Kent Bloomer notes that the term *ornament* is "loaded with conflicted and even negative meanings"; it has often been associated with notions of "excessiveness, frivolity, trivia, and superficiality" (15). Architectural ornaments themselves have passed in and out of favor for centuries. Their popularity is evident in the Ionic capitals of Greek temples, the rococo ornaments of the eighteenth century, and the Gothic

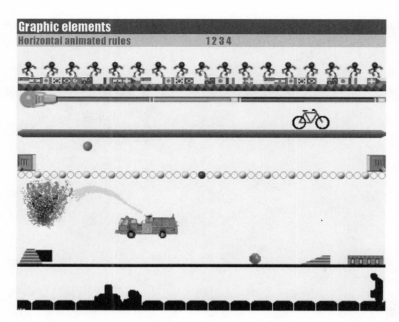

Figure 9.8. `<hr>`-like animated images.

revivalism of the nineteenth century; conversely, the modernist movements in twentieth-century arts, architecture, and design rejected elaborate ornamentation and strove for more minimalist representations.

E. H. Gombrich has pointed out that ornamentation has been attacked not just on aesthetic grounds, but also on moral grounds. Critics of ornament have argued that it "is dangerous precisely because it dazzles us and tempts the mind to submit without proper reflection" (17). Ornament has an almost seductive power; as Gombrich writes, "the warnings against displays of decoration are a tribute to its psychological attraction. We are asked to be on our guard because they may work only too well" (17). Those who argued against ornament preached the value of neoclassical simplicity and rationality—arguments that, as Gombrich points out, also apply to the realm of speech and rhetoric.

Even defenses of typographical ornamentation have been couched within the rhetoric of what Gombrich calls "the cult of restraint" (18). In 1849, A. W. Pugin, a proponent of Gothic ornamentation, wrote in *Floriated Ornament* that ornamentation must be approached in an "appropriate" manner:

Ornament, in the true and proper meaning of the word, signifies the embellishment of that which is in itself useful, in an appropriate manner. Yet, by a perversion of the term, it is frequently applied to mere enrichment, which deserves no other name than that of unmeaning detail, dictated by no rule but that of individual fancy and caprice. Every ornament, to deserve the name, must *possess an appropriate meaning and be introduced with an intelligent purpose* and on *reasonable grounds.* (qtd. in McLean 115)

Like David Siegel's attacks on the horizontal rule, Pugin's defense of ornament emphasizes the need for uniformity and utility. Central to his argument is a specific form of aesthetic taste, the measure of "an appropriate manner." Gombrich has suggested that an analogous controversy—attacks against the rococo style in eighteenth-century France—ended similarly with the requisite proffering of "justifications before the Courts of Reason and of Taste" (22).

The Web has helped collapse older distinctions between the crafts of writing and printing; with the creation of the Web, writers became printers, working not only on their texts, but also on the graphical frameworks used to display them. And with that change, the courts of reason and of taste have voiced displeasure with the democratization of design. In *Screen: Essays on Graphic Design, New Media, and Visual Culture,* Jessica Helfand has argued that the new technologies have bred an alarming "new illiteracy" that "presupposes a progressively libertarian view of the world, freed from the shackles of rulemaking and restrictiveness" (91). That libertarianism, Helfand argues, has resulted in a "sort of ill-defined, anything goes expressionism" (91) that is itself symptomatic of the need for a return to the principles of proper graphic design. She writes:

This is a new kind of illiteracy, encouraged by the widespread pluralism that colors contemporary culture on the one hand, and forgiven by the extraordinary impulse to self-publish that dominates it on the other. More often than not, the new illiteracy is an incubator for anger, a breeding ground for self-importance—the world according to me, me, me. Undisciplined in form and irreverent in tone, its denizens play by their own rules, invent their own taxonomies, and espouse their own highly individualized view of the universe. (92)

Helfand's conservative, reactionary polemic against this culture of "unbridled permissiveness" (94) has its roots in the neoclassical attacks on rococo design described by Gombrich. Like the critics Gombrich describes,

Helfand's objections to the newly "undisciplined" and egocentric culture she sees before her amounts to "no more than what psychologists call rationalizations of preferences fed from less conscious sources" (Gombrich 22). Helfand's call for a return to design principles betrays the anxieties of a profession whose authority is challenged by the rise of the newly empowered creators of content on the Web. And her scornful deprecation of those who would deign to "invent their own taxonomies" seems woefully out of date considering the important role that folksonomies have played in Web-based social networks.

The rules of the game, as Helfand notes, have been transformed. When we consider `<hr>`-like animated banners in light of these debates, we can better understand the value of serious considerations of tags such as the horizontal rule. They are, or should be, important to us not necessarily for their artistic value, but rather as indices of sociocultural values and attitudes. Kent Bloomer has argued that ornament itself is a performance that is both "combinational" and "transformative" in nature (27). The ornament is not just an adornment to a form, but an interaction with that form; it is "a habitat that allows metamorphosis" (27). The metamorphosis of the horizontal rule in the first decade of the World Wide Web is evidence of the new aesthetic styles, rooted in populism, that flourished as the Web began to influence popular culture.

On Line: Ruling Out Virtual Space

In "The End of Books," Robert Coover's 1992 meditation on hypertext for the *New York Times,* Coover famously noted that "much of the novel's alleged power is embedded in the line, that compulsory author-directed movement from the beginning of a sentence to its period, from the top of the page to the bottom, from the first page to the last." The World Wide Web promised to unsettle the hegemony of that line, and along with it traditional practices of reading and writing. The instantiation of virtual space could offer, Coover claimed, "true freedom from the tyranny of the line," thus bringing to fruition a series of experiments in narration over the centuries by authors ranging from Laurence Sterne to James Joyce. In the new medium, linearity would exist only if it was "implant[ed]" there.

Like the "conservative creatures" in Coover's writing classes who "frantically rebuil[t the] old structures" in new hypertextual spaces, early Web designers tended to plant linearity insistently in the uncharted space of the graphical Web. Some of the possibilities offered earlier in this essay—

that the horizontal rule was a remediated form of the rules that appeared in printed books; that the horizontal rule allowed designers to add visual elements to their pages without overloading the limited bandwidth connections of their users; that horizontal rules offered early Web designers a democratic and dramatic sense of play and creativity on otherwise spartan pages—suggest reasons why the horizontal rule was used so frequently in the early years of Web development, but they nevertheless point to the irony that one of the first things designers did with virtual space was to carve it up and fill it with lines and rules. Having encountered a medium that freed them from linearity and that seemed to demand an approach that would come close to what Derrida saw as "the end of linear writing" (86), designers hewed close to the strictures of older forms of media. It was precisely this situation that led David Siegel to call for the banishment of horizontal rules.

The question we face now is whether the banishing of the `<hr>` tag has in fact resulted in the banishing of horizontal rules. Put another way, although we may see the `<hr>` tag less frequently these days, horizontal rules still appear on Web pages in two forms. First, and most literally, lines formed with `<div>` tags and bottom borders continue to be used as dividers between sections of text. Second, and more importantly, the horizontal rule understood more abstractly—as the general injunction to write in a linear fashion, to compose pages that are read in linear ways— still dominates most reading and writing practices on the Web. For all of the multiple branchings of networked hypertexts, for all of the decentering potentiality of multimedia writing, popular Web-design trends continue to privilege the tenets of linearity and rationality over more disruptive organizational schemas.

Exactly what those alternate schemas might look like is an open question, but two recent developments show promise for continuing the kind of experimentation practiced by Coover and his students in the early 1990s. First, new forms of networked, distributed textual discourses, such as those that might be compiled through feeds of microblogging services such as Twitter or Identi.ca, have created new writing spaces in which individual textual units cannot be understood in isolation but rather must be combined with other textual units to form aggregated wholes.[2] In such environments, the act of aggregation itself becomes a form of writing, a form of textual inscription that reshapes meaning on the fly according to shifting sets of parameters.[3] In such spaces of aggregated discourse, traditional practices of linear writing and reading are disrupted by flows of data that continually reshape the space of the text.

Video-based communications would seem to hold out even greater possibilities for breaking the rules of linearity; as Lev Manovich has observed, there is "a general trend in modern society toward presenting more and more information in the form of time-based audiovisual moving image sequences, rather than as text" (78). And yet, as Jonathan Alexander and Jacqueline R. Rhodes have recently suggested, many attempts to incorporate video composition into academic classrooms tend to "bastardize" new media in that they merely transpose the logical structures of written rhetoric—the essay, the argument, the introduction and conclusion—into visual form. Alexander and Rhodes have delineated a new type of rhetorical pedagogy that takes for its model not the textual form of the traditional academic essay, but rather the poetics of avant-garde cinema. Pointing to the suggestive and associative visual transitions that link together sequences from films such as Jean Cocteau's *La Belle et La Bête* (1946), they have called for forms of narrative construction that are premised on a clean break with the old rules of linear composition.[4]

Such pedagogical and rhetorical movements represent a welcome embrace of the nonlinear aspects of virtual space. They hold out a promise that resembles what the pioneers of the Web sought for the new medium, which was a system of communication that would approach the operations of the human mind. Vannevar Bush's "memex," which would limn the forking paths of associative thought, mimicked not printed books but the process of thinking itself. Ted Nelson's notion that computers would "set us free from the dimensions and topology of paper" so that we could "find truer visualizations of the conceptual structure of what we do" (28) also sought to mimic not the linear structure of rational, paper-based argument, but the flitting glimpses of the mind in action.

As the Web continues to mature, future designers will decide which rules to follow and which to break. If they hope to follow the examples of Bush and Nelson, they might also consider the work of William James, whose theories of the mind bear a resemblance to Bush's memex. In *The Principles of Psychology,* James attempted to reframe dominant notions of the human mind:

> Consciousness, then, does not appear to itself chopped up in bits. Such words as "chain" or "train" do not describe it fitly as it presents itself in the first instance. It is nothing jointed; it flows. A "river" or a "stream"

are the metaphors by which it is most naturally described. *In talking of it hereafter, let us call it the stream of thought, of consciousness, or of subjective life.* (239)

Adapting Jamesian paradigms to new-media forms would entail the creation of graphical webs that would take as their grounding metaphors not the horizontal lines and rules of the printed text or the early Web, but rather the associative streams of thought itself. Perhaps we don't need to break or banish the rules; we just need to make them permeable, allowing streams of data to flow through them.

Notes

1. I am grateful to Bradley Dilger for this observation.

2. For helping me think through the implications of distributed textual discourse, I am indebted to John M. Jones for his presentation in the "Microblogging: Producing Discourse in 140 Characters or Less" session at the 2009 Modern Language Association convention.

3. The notion of writing spaces created on the fly holds certain resemblances to what David Harvey has called the principle of "flexible accumulation" in postindustrial capitalist societies. Harvey writes that flexible accumulation is characterized by "enhanced flexibility and mobility" that have "greatly intensified rates of commercial, technological, and organizational innovation" (147). In spaces of textual aggregation, textual production and consumption are compiled on demand and change according to the preferences of individual users. As Lev Manovich has noted, "If the logic of old media corresponded to the logic of industrial mass society, the logic of new media fits the logic of the postindustrial society, which values individuality over conformity" (41).

4. Mark Sample has recently called for a similar reimagining of electronic books, noting that "the real problem isn't so much that e-readers won't let me read books that experiment with form. The real problem is that most novelists are writing books that don't experiment with form."

Works Cited

Alexander, Jonathan, and Jacqueline R. Rhodes. "An Illegitimate Profession: How Composition Bastardizes New Media Scholarship." Presented at the 124th Annual Convention of the Modern Language Association, San Francisco, Calif., 26–30 December 2008.

Arnold, Edmund C. *Functional Newspaper Design.* New York: Harper & Brothers, 1956.

Barnhurst, Kevin G. *Seeing the Newspaper.* New York: St. Martin's Press, 1994.

Berners-Lee, Tim, and Daniel W. Connolly. "Hypertext Markup Language—2.0." Network Working Group, MIT/W3C, September 1995. Accessed 15 November 2006. http://www.w3.org/.

Bloomer, Kent. *The Nature of Ornament: Rhythm and Metamorphosis in Architecture.* New York: Norton, 2000.

Bolter, Jay David. *Writing Space: Computers, Hypertext, and the Remediation of Print.* 2nd ed. Mahwah, N.J.: Lawrence Erlbaum, 2001.

Bolter, Jay David, and Richard Grusin. *Remediation: Understanding New Media.* Cambridge, Mass.: MIT Press, 1998.

Bush, Vannevar. "As We May Think." *Atlantic.* July 1945. Accessed 15 January 2009. http://www.theatlantic.com/.

Coover, Robert. "The End of Books." *New York Times.* 21 June 1992. Accessed 14 January 2009. http://www.nytimes.com/.

Derrida, Jacques. *Of Grammatology.* Translated by Gayatri Chakravorty Spivak. Corrected edition. Baltimore: Johns Hopkins University Press, 1997.

Gombrich, E. H. *The Sense of Order: A Study in the Psychology of Decorative Art.* Ithaca, N.Y.: Cornell University Press, 1979.

Harvey, David. *The Condition of Postmodernity: An Enquiry into the Origins of Cultural Change.* Cambridge, Mass.: Blackwell, 1990.

Helfand, Jessica. *Screen: Essays on Graphic Design, New Media, and Visual Culture.* New York: Princeton Architectural Press, 2001.

Hickson, Ian, ed. "HTML5: A Vocabulary and Associated APIs for HTML and XHTML. W3C Working Draft." 24 June 2010. Accessed 7 August 2010. http://www.w3.org/

Hutt, Allen. *The Changing Newspaper: Typographic Trends in Britain and America, 1622–1972.* London: Gordon Fraser, 1973.

Huttner, Sid. "R. Worthington & Co." *The Lucille Project.* Accessed 5 March 2007. http://sdrc.lib.uiowa.edu/lucile/.

James, William. *The Principles of Psychology.* Vol. 1. 1890. New York: Dover, 1950.

Jacobs, Ian, et al, eds. "HTML 4.01 Specification." *W3C.* Accessed 3 March 2007. http://www.w3.org/.

Jones, John M., "Twitter at MLA II: Panel Notes." *HASTAC.* 5 January 2009. Accessed 15 January 2009. http://www.hastac.org/.

Landow, George. *Hypertext 3.0: Critical Theory and New Media in an Era of Globalization.* Baltimore: Johns Hopkins University Press, 2006.

Lipson, Carol S. "Recovering the Multimedia History of Writing in the Public Texts of Ancient Egypt." In *Eloquent Images: Word and Image in the Age of New Media,* edited by Mary E. Hocks and Michelle R. Kendrick, 89–115. Cambridge, Mass.: MIT Press, 2003.

Luidl, Phillip, and Helmut Huber. *Typographical Ornaments.* Poole, Dorset: Blandford Press, 1985.

Manovich, Lev. *The Language of New Media.* Cambridge, Mass.: MIT Press, 2001.

McLean, Ruari. *Victorian Book Design and Colour Printing.* Berkeley: University of California Press, 1972.

Meggs, Phillip B., and Alston W. Purvis. *Meggs' History of Graphic Design.* 4th ed. New York: John Wiley & Sons, 2006.

Mitchell, W. J. T. *Picture Theory: Essays on Verbal and Visual Representation.* Chicago: University of Chicago Press, 1995.

Nelson, Ted. *Dream Machines/Computer Lib*. Redmond, Wash.: Tempus Books of Microsoft Press, 1987.

Reilly, Elizabeth Carroll. *A Dictionary of Colonial American Printers' Ornaments and Illustrations*. Worcester, Mass.: American Antiquarian Society, 1975.

Ryder, John. *Flowers and Flourishes: Including a Newly Annotated Edition of "A Suite of Fleurons."* London: Bodley Head for Mackays, 1976.

Sample, Mark L. "Do You Like Your E-Reader? Six Takes from Academics." *Chronicle of Higher Education*. 13 June 2010. Accessed 7 August 2010. http://chronicle.com.

Siegel, David. *Creating Killer Websites: The Art of Third-Generation Site Design*. 2nd ed. Indianapolis: Hayden Books, 1997.

Woodman, Mark. "Horizontal Rule with a Spinning Ball in the Center!" *Adveractive*. Accessed 7 March 2007. http://adveractive.com/.

10 BODY ON <body>

Coding Subjectivity

JENNIFER L. BAY

<body>, *an HTML and XHTML container element, encloses the content of a page: links, images, text, and other included features. Every page must contain a* <body> *tag, and with few exceptions, content outside the* <body /> *container is not rendered by the browser, though it may inform behavior or affect style. Early Web pages did not use* <body>*. The tag was not required until HTML was more formalized with HTML 2, reflecting the rapid growth of the complexity of HTML and a sense that better organization of the code was needed. The Web browser Mosaic, for instance, did not include support for* <body> *until July 1995, almost two years after the browser was released. As the number of Web sites grew geometrically, the number of methods, opportunities, and approaches for constructing such sites grew as well. The simple division of HTML documents into two parts—head and body—afforded HTML the gravitas of HTML's predecessor, SGML (Standard Generalized Markup Language), and presented Web writers with a clearly defined space for work.*

THERE IS NO LONGER ANY DOUBT that bodies inhabit, produce, and function as code. Whether genetic, semiotic, or mechanic, the body is constructed and interpreted as a series of complexly interactive codes that contribute to the creation of what we might call a subject. Similarly, the <body> tag in Hypertext Markup Language assists in the creation of what we might call a subject on the Web. The <body> tag operates at the semantic and local levels of HTML (or XHTML) to demarcate the content of a Web page. It functions as a boundary within the code, indicating where the content of the page begins and ends, much like the skin of the human body is often the perceived boundary between self and world. Other tags, Cascading Style Sheets (CSS), and scripts can define how that content or body is presented online (for example, color, font, JavaScript). In short, the <body>'s attributes—the look and presentation of the body—are often determined by other tags. Both body and `<body>` are marked up, and it is in the act of the markup that subjectivity emerges.

The concept of marking up or tagging bodies inevitably brings with it the valence of cultural markers and the role that they play in how the body is both interpreted and inhabited. Early Internet research heralded the potential to escape the body and its markers (for example, Gibson; Bruckman; Rheingold). Megan Boler, among others, has argued that although we once thought that the Internet would allow users to transcend the social and political implications of their bodies, in fact the opposite has held true. She explains, "Despite the hypes and hopes of the freedom offered by transcending usual images of the other, there comes a point at which users crave information about traditional markers of the body" (140). Boler provides various examples of online interactions to show that often "users require that the other offer 'essential' data about their 'real life' identity so that sense can be made of textual utterances" (153–54). Throughout most of the studies on online interaction, Boler observes that the body serves as the final arbiter of authentic identity (157). The revealing of these real bodies has implications for the rhetorical encounters that happen online. Obviously, as Boler herself notes, when we share essential data about our bodies, we also risk bringing our cultural prejudices into our online encounters. Moreover, such information also affects the ways in which we comport ourselves toward others. While there are plenty of crucial ways that we and others tag our bodies online, gender is crucial for how we position our bodies, and those bodies are positioned by code. The body as a key arbiter of identity constitutes and constrains the types of rhetorical interactions available to both women and men online.

In practical terms, we know from reports such as the Pew Internet and American Life Project that overall, women use the Internet slightly less than men, and they use it somewhat differently, most notably for social connection rather than information access. However, women under thirty actually outpace their male counterparts in Internet usage. How, then, are these gendered, embodied subjects who use the Web reflected in the tags that define its content? For instance, because the `<body>` tag indicates the place where most content of a Web page resides and will thus show up for readers, we might say that meaning lies in the potential of that "body." This reversal of the Cartesian dualism allows us to rethink how material codes are both reflected in and affected by the codes that define the content of the Web.[1] How does what is inside that `<body>` tag physically and psychically position us as human beings? That is, how does the `<body>` tag code our subjectivity?

The Amorphous `<body>`

The `<body>` tag is a site of incredible potentiality. Although other HTML tags might be seen as more influential (the `<html>` tag, for instance, allows a Web site to actually appear), the `<body>` tag functions as a repository for a Web page's content. Without the `<body>` tag, or without anything within that tag, the page appears empty until the writer adds textual or visual content. The body of a Web page can be, can contain, anything; online, the body is infinitely malleable. What the site "does" is usually found in the meta tags before the `<body>` tag—they instruct what the body can do, what is possible. This amorphous body is in line with postmodern feminist discussions by writers such as Elizabeth Grosz, who seeks to rethink subjectivity from the vantage point of corporeality. Although Grosz by no means argues that the body is a blank slate to be coded or inscribed, she does see the body as containing an infinite potential for inscription and reinscription. Such inscriptions form much of the basis for how sexuality and gender are coded on human bodies. But such inscriptions are not windows for a psychic interior. Rather, Grosz is more drawn to a Deleuzian sense of "the body as a discontinuous, nontotalizable series of processes, organs, flows, energies, corporeal substances and incorporeal events, speeds, and durations" (164). In such a model, the body does not function within a dualistic hierarchy (interior/exterior or mind/body), but emerges through specific practices.

Although Grosz's work does not address the Internet or cyberspace, instantiations of `<body>` on the Web are equally as applicable to her theories. The last five years have seen a plethora of articles and books that attempt to embody materiality within code and technology. Two of the most prominent theorists, N. Katherine Hayles and Mark B. N. Hansen, posit that material bodies are present in new media works and the software that compose them. Although this focus on "coding materiality" is significant, what has been lacking from those discussions is a clear sense of how subjectivity is coded within everyday Internet communications. That is, most of the examples used in Hansen and Hayles focus on new media installations and artworks, not on the everyday Web pages that most human beings read and write. This chapter theorizes how embodied subjects come into being through computer-generated and human-programmed code on the Web.[2]

Although Hansen and Hayles usefully demonstrate the materiality of bodies in new media artworks, theories more grounded in feminism and gender theory give us better access to the lived experiences of human bodies online. For instance, in her book *Getting under the Skin: Body and Media Theory*, Bernadette Wegenstein uses a feminist lens to read the history of the body as a "revealing of the *body as mediation*" (35). In this paradigm, mediation is not something that "just happens" to a preexisting body; rather, mediation is the process through which bodies emerge as such, in all their material glory. As Wegenstein goes on to explain, the rise of body criticism over the past twenty-five years is, not coincidentally, simultaneous with the increasing medialization of society. Arising from the tension between a concept of the body as holistic and a concept of the body as fragmented, Wegenstein shows that "what has previously been known as *medium* has adopted the characteristics of *body* within the techno- and new-media sphere of the new millennium" (120). Using examples from new media installations, advertisements, architecture, and performance art, she demonstrates that "both fragmentation and holism are indispensable modes of imagining and configuring the body" (3). One of these ways of configuring the body is the synecdochic move of having any part stand in for the body. Wegenstein explains, "The face, which has always 'overcoded' other body parts, has now ceased to be the most representative signifier of human appearance; under the skin every organ has an (inter)face. Potentially, each organ may stand in for the whole body" (80). This is particularly the case with gendered representations in which sexed body parts come to stand in for the entire body. Moreover, Wegenstein continues to explain that in a world where one part can overcode the rest of the body, the body itself seems no longer necessary. Rather, in a play on Deleuze and Guattari's body without organs (BwO), she insists on "'organs instead of bodies (OiB),' namely organs that are configured as inside out, having lost their quality of being 'in' the body" (80). In this instantiation, the OiB is a "'flattened' body that has attained the value of a screen, a surface of reflection—in other words, a medium itself" (80).

Examples abound in popular culture, where the flattened body or the skin in particular has become a medium. Perhaps the most obvious instance is tattooing, which has gained increasing mainstream interest in the past twenty years; one figure cites that 50 percent of twenty-one- to thirty-two-year-olds in the United States bear tattoos. In one example, Wegenstein discusses a Jenny Holzer advertisement for the designer Helmut Lang's new perfume line that highlights the skin as surface: "On my skin I smell you I breathe you I tease you I wait for you I scan you I watch you I see you I walk in" (111). Wegenstein observes that in the ad, "the skin's surface does not only refer to a physical experience; rather, the skin is the metonymical platform for any human contact" (112). In short, the largest and flattest organ, the skin, no longer functions as the body's boundary, the subject's boundary; rather, it has itself become a medium. Wegenstein explains, "The skin opens up various spheres, physical and beyond, for human encounters" (112). Although the skin takes prominence in this instance, in an OiB world, any organ can substitute for the body. Discrete parts of the body become the media by which communication and connection can take place.

Like Hansen and Hayles, Wegenstein is drawn toward artistic interpretations or readings of the phenomena she analyzes in new media art. However, her "notion of *body as constitutive mediation*" (120) is more germane to the complexities of the Web as both a disembodied and embodied site. Some of our everyday encounters on the Web serve as examples of this phenomenon of constitutive mediation, especially in Web sites that specifically (allow us to) construct our bodies. Using attributes, the `<body>` tag can be styled in ways that determine the appearance of information on a Web page. We can see that Cascading Style Sheets, for example, function in some ways like the clothes we wear on our own human bodies. CSS defines the look and feel of a Web site—telling the content how to look on screen, what should be emphasized, and so on. Similarly, human bodies are coded by other attributes in Web spaces. We are constantly asked to provide evidence of our material bodies in Web space—where we live, what we look like, who we are—further demonstrating how the body functions as mediation.

In the remainder of this chapter, I look at what I see as three different kinds of Web spaces in which bodies are "tagged" and take on mediative properties that construct subjects. I use the term "body mediation sites" (or body delivery systems) because it is clear that the body is the constitutive form of mediation in these spaces. The three kinds of sites I discuss are dating sites (such as Match.com), social networking sites (such as Facebook), and blogs (such as anonymous/pseudonymous blogs). Each site allows for certain kinds of cultural codes, which are invented and arranged by computer code and which function as the attributes or markup of the body. Such codes affect the kinds of rhetorical encounters available online, as well as the comportment of bodies toward one another. For example, social networking sites elicit a nodular bodily comportment; you are a node in a network and (many) connections are what you are afforded. Dating sites require the subject to seek one other person; the subject is searching for one "match." Blogs can embody a more declamatory stance; the subject seeks to present ideas, whether for oneself or for others. In short, these sites function to help construct certain types of rhetorical or meditative bodies, but not in the discursive or textual manner most rhetorical criticism pursues. Rather, these sites provide the means, sometimes rigidly so, for the body as mediation to show up for us.[3]

Sites of Body Mediation

My reliance on a theory of body as mediation challenges previous Internet identity research, primarily because those theories were all too often entrenched in the embodied/disembodied binary. Along with postmodern theorists such as Donna Haraway and Jean Baudrillard, Sherry Turkle, in her early landmark text, *Life on the Screen*, highlighted the fragmentation of the self and the ability for subjects to inhabit multiple identities in online environments. Before the Web explosion of the late 1990s, most online interaction occurred through e-mail, chat, Usenet, and textual games and social spaces such as MOOs (multiple-user, object-oriented environments). In many of these spaces, users assumed and played with multiple identities. Although we would all acknowledge that the ability to play with identity online remains, essays such as Julian Dibbell's "A Rape in Cyberspace" was one of the first to call into question the connection between online and off-line selves. That essay brought with it much debate over whether the online actions detailed by Dibbell constituted rape, as well as what kinds of consequences were experienced by the real-life women involved in the situation. More than ten years later, the commercialization of the Web has pushed online and off-line selves to become much more continuous, making some of the issues Dibbell first addressed appear archaic.[4]

What we do online now requires there to be more continuity—or at least more fluidity—between our online and off-line selves. We pay bills, communicate volumes through e-mail, buy and sell, perform research to benefit our material lives (such as home improvement), manage money and financial accounts, request information, hook up utilities, and more— all of which are connected to our material, off-line selves. Even in 1999,

Lori Kendall argued that "on-line interaction cannot be divorced from the off-line social and political contexts within which participants live their daily lives" (58). Indeed, she points out that although we may see more evidence of identity multiplicity online, people engaged in different presentations of self to different audiences long before the Internet (61). We understand this, of course, as rhetoric. It's true that identity play is allowed quite a bit in gaming environments but appears less frequently in the "real life" of the Internet, and as I will show, there is still an extreme desire for users to know that others are authentic, real bodies. As Internet use has become mainstream and commercial, this assumption—and desire for—online and off-line selves to be one and the same may have emerged from individual interactions, but it is facilitated and constructed by corporate America and corporate-funded Web sites. Such facilitation functions to constrain the types of bodies available for presentation online, as well as the types of rhetorical interactions. These interactions—and bodies—are also dependent on the role of fantasy, further demonstrating how the body (or its parts) has become a screen, has become mediation.

In his article, "'Quality Singles': Internet Dating and the Work of Fantasy," Adam Arvidsson argues that informational capitalism has activated, to an unprecedented amount, what he calls "'fantasy work'—the work of imagining situations, people and relations" (672). Arvidsson observes that capitalist enterprises on the Web, particularly in dating sites, commodify affect in order to create bonds and meaningful interactions. I will return to Arvidsson's discussion of dating sites such at Match.com. For now, I want to point out that his concept can be extrapolated to other parts of the Internet in our daily encounters with others. We want to believe that those we encounter are authentic—that our encounters are essentially human. Social networking sites are perfect examples; such sites allow one to accumulate contacts, friends, or colleagues, making one feel as if he or she has many real-life connections. We even bring our beliefs, attitudes, and behaviors online and project those onto our encounters online; we saw this in the 1990s in both Sandy Stone's and Sherry Turkle's discussions of instances of Internet deception, and we still can see it today. Take, for instance, the outrage at the 2001 Kaycee Nicole blogging hoax. Her Internet audience wanted to believe in this young woman's plight, and they formed an affective attachment to her narrative. Readers invested in her story; they wanted to believe she was real, just as we can see in the media interest from true-life stories such as Armistead Maupin's *The Night Listener,* in which an adult impersonates a young survivor of incest struggling with AIDS.

Wegenstein interprets this desire to get to an authentic reality as a desire to get beyond the screen. She observes that as more ways of representation are created by new media, "the desire for the real and for the 'authentic first person's experience'" grows stronger (70). The obvious way to get away from representation is to remove the screen, and one way to do this is "to (make) believe that there is no screen, or that the medium is not a medium but in fact part of the real" (72). Another approach, Wegenstein explains, is to *"make the spectators aware of being spectators, make them enjoy the medium!"* (72–73). Reality TV shows where audience members participate are just one example of this phenomenon. The Web abounds with multiple examples of the desire for the authentic first-person perspective, especially during and after crises such as the Virginia Tech shootings or the 11 September 2001 attacks. This desire for "real" human encounters is essential to constructing the fantasy of authenticity online.

In an environment where identity is so completely data driven, material markers are one of the only ways to ensure meaningful human interaction. Because one's identity is coded by pertinent information, we can build a body from this information. Isn't this how identity theft works? One steals a person's vital information—the vital code that identifies them—and uses that information to present oneself as another. One is reminded of the MasterCard television commercials in which the body of the identity theft victim speaks with the thief's voice and participates in activities that are incongruous with the victim's body. In these cases, the real body can be the ultimate clue to identity, but it can also be considered suspect, which is why so many people who are victims of identity theft have such a difficult time convincing authorities of their authenticity.[5] We might theorize this phenomenon by saying that if bodies are mediation, then early Internet research indicated a desire to remove the screen, remove the body, and achieve a more equitable reality. Today, the extreme consequences of mediation, such as identity theft, have caused us to move to the opposite direction and get beyond the screen to find more and more "authentic" markers of the real body.

Plenty of Web sites strive to get beyond the screen by attempting to present real bodies as much as possible and creating community around those attempts. The early genre of the Webcam was one early way for individuals to present and access real bodies on the Internet (see White). The Web 2.0 phenomenon has been characterized by its focus on connection and on the network. I would extrapolate this to say that it is a connection among bodies. Social networking sites allow individuals to network or connect with others on the Web, be it around professional issues, common interests, or education. Facebook is one of the most prominent examples, and as the name implies, the site is a collection of "faces," or individuals who want to connect with others around common experiences, most of which are aligned with a material space such as a school or place of work. The site bills itself as "a social utility that connects you with the people around you." Facebook, like other social networking sites, operates as a closed community; in order to belong, one must register for the site. Although an extremely limited profile listing appears in a Google search, the community is only searchable by members. "Everyone can join," but the site encourages an affiliation with a real, material site—a particular high school, college, location, or company. This material connection sets the stage for online interactions to occur. Implied in this arrangement is that you will connect with real-life friends or acquaintances connected with that material space.[6]

Members develop a profile, of which a self-image is a key indicator of identity. Members are asked to complete a profile with "your photo and interests, your work and education history, your favorites and more." Although many individuals choose photographs of their faces (or their entire bodies) for their profile pictures, as many others do not. This returns us to Wegenstein's idea that in a world where the body is the constitutive medium, "any body part can become a face" (97). I would extrapolate that to say that any thing can become a face, or in this case, a profile—a particularly angled view of the body. If body parts are becoming separated from their natural body environments, as Wegenstein suggests, then other things can become separated from their environments and can stand in for the body. Members can choose cartoon characters, pets, or other objects, and they can change their profile pictures at any time; such images "body" the identity of the member within the network. Moreover, the default image for members without a profile picture is a faceless white head on a blue background. The outline of the head is gender unspecific, but this unspecificity tends toward a masculine look (short hair, collared shirt). The lack of image makes it especially difficult to determine whether this is indeed someone you would like to be friends with. Members also develop textual profiles that contain what I would argue function as attributes— favorite TV shows, quotations, birth dates, education history, sexual orientation. All of these attributes, especially orientation, construct a comportment for the member that directs the body toward shared connections with others. Indeed, only friends that you choose can see your profile. One feature of Facebook where we see this most explicitly is the "wall," a site where members can leave notes. Writing on one's wall becomes a sort of writing on the body; the wall takes on a skinlike boundary, and users can write and respond on each other's walls.

Other attributes that form the "body" of the individual include the status feature, which allows the individual to indicate what is currently happening. The default syntax of the text used to be the "to be" verb: "User X is _____," implying that the individual is actively doing or being. The "to be" verb has been removed as the default syntax, but there is still the expectation of an active doing or being on the part of the user. This kind of ontological valence also emerges in the news feed that dominates a user's main entry page. This feed is a rolling list of changes and updates that your friends have made to their profiles. Not only do you see what others have been doing, but your updates are also listed in their news feeds. Facebook faced a backlash from users when they rolled out this feature in early 2007, mainly because of the heavy traffic they would face when users often have hundreds of friends; rather than succumb to the hostility of their users, they refused to remove the feature and only allowed users to opt out. One Facebook executive was even cited as saying that users will "get over it," implying that Facebook will do things as they see fit, and providing further evidence that Facebook assumes and is pushing users to have real bodies and make real connections. But real bodies and real connections have consequences, and as the network grew increasingly large in 2009, Facebook users started to realize that there were privacy concerns with sharing information with hundreds of "friends." Today, there are numerous Web sites that explain how to adjust your privacy settings on Facebook to limit how much information you want to share. The discussion around privacy concerns, however, is another acknowledgement that Facebook assumes and wants users to match up with their physical, real-life identities.

The material constraints that Facebook imposes in order to build the fantasy of an authentic body of friends become most obvious with their policy on false profiles. There are a number of fake profiles on Facebook; in fact, the Web site Fake Your Space creates fake friends for you on Facebook or MySpace to increase your connections and popularity. However, Facebook discourages such fakes and asks members to report them, adding to the pressure to "be yourself" and represent your real-life self as accurately as possible. This further pushes members to be comported toward others in a nodularlike capacity, encouraging members to believe not only the fantasy that such connections are real and meaningful, but that a member's profile is equated with the true body of the member. As I mentioned earlier, identifying individuals without pictures is difficult. Only those with real bodies can connect within the network.

Arvidsson points to this same pressure to represent one's real self as accurately as possible on Internet dating sites. Although he notes that such dating sites are "places where the powers of fantasy are stimulated" (678), he also observes the push toward the presentation of an authentic self, which he sees as coming mostly from the corporate sites themselves. Arvidsson writes, "Communication and interaction on Match.com evolves on the assumption that true love is contingent on a true and authentic experience of selfhood, on revealing the inner self and its true desires" (682). But that authentic self is also dependent on the presentation of the body; as Arvidsson observes, Match.com "proposes that body type mirrors the person's inner true self and that true love with a compatible partner is contingent on a truthful presentation of these qualities" (684). He provides several examples from the Match.com "12 Tips to Pen Perfect Profiles" that encourage members to present their bodies in pictures in acceptable ways. In this example, the comportment of the body is essential to the construction of the subject. Through the construction of a particular kind of profile, the member presents him- or herself as an authentic real-life body that corresponds to what Arvidsson describes as a "quality single" ideal perpetuated by Match.com.

Since Arvidsson conducted his research on Match.com in 1995, the site has shifted slightly in the advice it provides to members seeking to develop profiles. The site now uses the term *portraits* to describe member profiles, further emphasizing that members are not just creating a side view of one's body—implied by the term *profile*—but a "real" image of one's real body. In the help section, one can still find numerous older articles to assist members in constructing profiles, but the article "12 Tips to Pen Perfect Profiles" is no longer up. Instead, the help feature is now dominated by a series of videos called "The Voyeur's Guide to Match.com," featuring Jay Manuel, a fashion photographer famous for his work on TV reality show *America's Next Top Model*. In these videos, Manuel goes through several "successful" member portraits and then helps three members overhaul aspects of their portraits. Along the way, he gives advice for creating a portrait, including tips such as "discover the real you" and "summon your true you" in photos and in words. The tagline to the guide is "It's ok to look," which explodes any remaining idea that the site is performing anything less than fantasy work. Manuel's presence is crucial because, as Arvids-son reports, the majority of members are women (681). Female members are led through the fantasy of being styled and counseled by a staff member of *America's Next Top Model,* implying that they can be such models (or that they should develop a "model" countenance). Also critically apparent is the fact that Manuel works in the modeling industry, where he produces images and deals with female bodies. Thus, what becomes apparent is that like on Facebook, members must develop a particular bodily countenance online in order to connect, but unlike Facebook, in which one body functions as a node in the network, Match.com members comport their bodies toward one other particular type of body in search for a connection. Therefore, the body becomes much more structured and managed in this type of Web site; one's attributes determine whether a match can or will take place.

Moreover, there is no potential for a false portrait on Match.com. A fake portrait on the site disrupts its rhetorical foundations. Although funny avatars or profiles on Facebook work because of the networked orientation of the site, in one-to-one encounters such as Match.com, even misrepresenting oneself fails. The expectation of the real, exact body is necessary for the fantasy to work, and in this case, the fantasy upholds the rhetorical situation.

Perhaps one reason that we see constraints on how bodies function as media in these spaces is that corporate America sees the Web first and foremost as a delivery system for information, for commerce. Most commercial Web sites operate on a broadcast model, and one of the newest corporate models is to exploit the social relationships that emerge from those networks (for example, see Locke, Searls, and Weinberger's *The Cluetrain Manifesto*). When those corporate sites are predicated on making connections to the network, bodies often become the media of the site. Unexplored so far are the reasons why corporate sites build in constraints that force users to inhabit real bodies. Part of the reason, of course, is that those material traces are needed to enhance the fantasy of connection/human contact the sites purport. But it also feeds into the corporate strategy for advertising. Corporations need real people—real-life consumers—to pitch their ads, a key revenue generator for most sites. Bodies, in this sense, become pure mediation. But bodies also function as media through noncorporate sites such as blogs. Although blogs can be very one-directional in the broadcast model format, they can also be collaborative and conversational. The `<body>` tag of a blog page reminds us that there are bodies writing the text and reading the page; without input from the blogger, the body of the page would be empty.

Several of the more unique features of a blog serve to form what we might see as an accessible body. Because a blog is an aggregation of writings by a writer or a series of writers, archives of past posts can show how a blog developed over time. In a study on accountability and blogging, Fernanda B. Viegas explains that blogs are archive oriented: "By allowing readers to examine earlier postings at will, blogs provide an informative backdrop against which to situate more recent writings." In this way, she continues, "regular readers thus can get a sense of the identifying 'voice' behind the posts on the site." This emphasis on voice not only implies that there is a human being behind the writing, but it also indicates that there is a body writing the text. Viegas even uses the term *portrait* to refer to how a blog archive forms an evolving identity of a writer. Unlike the portraits on a site such as Match.com, a blog portrait is accessible and available to anyone. Indeed, the rhetorical orientation of a blog is so open that Viegas sees this as a liability to be overcome. The study she conducted "suggests that authors are not overly concerned with preventing their genuine identities from being disclosed on their blogs." In addition to this openness, blog writers often do not know their audiences, which she sees as a potential problem that can result in unanticipated readers.

We might rethink Viegas's conclusions through a consideration of Helen Kennedy's research on online anonymity. In her article "Beyond Anonymity, or Future Directions for Internet Identity Research," Kennedy provides a useful perspective on the issue of anonymity by pointing out that there is a distinction between feeling anonymous and being anonymous, but that this distinction has not been addressed in the research on Internet identity formation. Her study followed several women from a variety of ethnic backgrounds who constructed and maintained Web spaces. She reports that for some of those women, the distinction between being anonymous and feeling anonymous was blurred. These women would post personal information and intimate details of their lives online because they felt like they were anonymous on the Web, even though they were not writing anonymously. Anonymity as an affect gets back to Arvidsson's idea that we engage in fantasy work online, and this kind of work is no less prevalent in blogging circles than in corporate-sponsored communities. Writers often buy into the fantasy that one loses his or her body—one's material accountability—online. But in fact, perhaps what happens is the body is fantastically produced by one's readers. Although we might think that the writer's archive produces a body of work, it is in the reading of those posts that a body emerges. In short, blogging—another form of media—is simultaneously another form of `<body>` and body.

We can further position this issue with respect to our need for a real, human, identifiable body beyond the screen. Although a blogger's "body" gets constructed within the texts he or she produces over a period of time, many attributes develop that body, and for some readers, there is the need for a proper name to point to the real-world body of the writer in order to establish ethos. This has resulted in many debates over anonymous or pseudonymous blogging practices. Research, along with anecdotal evidence, points out that more female bloggers are likely to practice anonymous or pseudonymous blogging. Critics of this practice argue that those who blog anonymously should be less respected and valued than those who blog under their real names. Somehow, these individuals are not taking responsibility for their words or actions because they aren't willing to reveal their material, off-line identities. Others respond that anonymity allows bloggers to explore personae and writing from perspectives that they might not ordinarily. We can connect this clearly to feminist theories of the body and to rhetorical comportment. If blogging invites what we might call an open comportment, meaning that a writer must comport him- or herself to anyone who encounters that blog, then it would make sense that women might feel vulnerable in such an open rhetorical encounter. And although such writers might feel anonymous, they often conflate the affective quality of feeling anonymous with being anonymous.

Conclusion

If bodies have become mediation, then what does this do to the concept of subjectivity? Certainly, we have almost always understood subjectivity as mediated, but it also brings us back to the literal concept that bodies are technologies, are media. Subjects in this sense are always technological, always coded, and always bodied. When we engage in rhetorical encounters online, the technologies we use constrain the kinds of bodies we can inhabit and thus constrain the ways we can interact with others online. For women seeking connection in the Web 2.0 world, this has gendered implications; while newer networking technologies seem to have the potential to create community and escape traditional hierarchical constructions, the drive for real identity disadvantages women from being accepted as authentic or expert (blogging), constrains the types of bodies and connections that can be made in gendered terms (social networking and dating sites), and affects the ways that women can comport themselves toward others. In other words, the comportments afforded for the bodies on these sites trigger gender codes that affect the possibilities for rhetorical encounters.

One of the driving forces of Kennedy's claims about anonymity is "to call into question the usefulness of the concept of identity" (861). Although calling the concept into question can be problematic for disenfranchised groups, she encourages such questioning "without losing sight of identity as embodied experience, of the real struggles of real people whose identities are fiercely contested or defended—in other words, without losing sight of identity-as-practice" (873). We might rethink this idea of identity as practice in terms of bodies and mediation as body as practice; that is, bodies in the Web sites I have discussed are mediations, and they emerge through specific online comportments, constraints, and practices. What this conception affords us is an understanding of the body as rhetorically involved in online communication, both within and beyond code.

Notes

1. I am not arguing for a reversal of the Cartesian dualism. In fact, I use a materialist feminist conception of the body that elides essentialist discourses about gender but that acknowledges the fundamental importance of the body in the creation of subjectivity. I posit such a reversal to examine what shows up when we look at the binary from the opposite end.

2. There has been a lot of research on how bodies emerged in online interactions before or at the beginning of the Web explosion. Much of this work centers on how traces of the body show up in text-based interchanges in educational settings or massively multiplayer online role-playing games. Less work has been done to address how bodies emerge through contemporary Web communications.

3. Absent from my examination is one of the most obvious sites where bodies take precedence: pornography. In some ways, pornography is the ultimate example of body as mediation, but I do not look at these sites for a variety of reasons. One reason is that I'm unfamiliar with them and with the overall genre, and another is that I believe that such sites are more geared toward viewing/looking than interaction. The examples I will discuss all provide space for bodies to interact with one another, whereas most porn sites are one-to-many delivery systems (rather than many-to-many systems). I'm sure there are some pornographic sites that do allow for interaction or fit into my taxonomy, but because the genre is arguably the most prolific and profitable on the Web, it necessitates its own discussion elsewhere.

4. Such issues still seem prominent in virtual worlds and online gaming environments. For instance, in the massively multiplayer online role-playing game *World of Warcraft*, there was a 2006 controversy about the ethics of a horde attack during the

online funeral for a popular player. A horde group ambushed and killed alliance players attending the online memorial service for a fellow horde character. The player had actually passed away in real life, and discussion abounded online about the ethics of the raid during a solemn memorial service.

5. An article in *Good Housekeeping* alerts readers to a new form of identity theft in which an individual steals another's identity for medical bills and reimbursement (Engeler). The article details one instance in which a victim's entire medical history was replaced with the thief's information, and the woman's children were almost removed from her home because of the action of the thief. It took the victim quite a while to convince the authorities that she had been a victim of identity theft. In another instance, a young child's Social Security number was stolen and used by numerous individuals, ruining her credit; the Social Security office refuses to assign her a new number, further linking one's body and data.

6. Until recently, there was an option not to be connected with a network. If chosen, an individual's profile was tagged with "no network," almost implying that the person was separated from others, floating alone in Facebook. Today, users are asked to affiliate themselves with their high school, college, city, or workplace, although that step can be skipped during registration.

Works Cited

Arvidsson, Adam. "'Quality Singles': Internet Dating and the Work of Fantasy." *New Media and Society* 8, no. 4 (2006): 671–90.

Baudrillard, Jean. *Simulacra and Simulation.* Translated by Sheila Faria Glaser. Ann Arbor: University of Michigan Press, 1995.

Boler, Megan. "Hypes, Hopes and Actualities: New Digital Cartesianism and Bodies in Cyberspace." *New Media and Society* 9, no. 1 (2007): 139–68.

Bruckman, Amy. "Gender Swapping on the Internet." *Proceedings of INET '93.* Reston, Va.: The Internet Society, 1993. Accessed 15 August 2007. http://www.cc.gatech.edu/.

Deleuze, Gilles, and Félix Guattari. *Anti-Oedipus: Capitalism and Schizophrenia.* Translated by Robert Hurley, Mark Seem, and Helen R. Lane. Minneapolis: University of Minnesota Press, 1983.

Dibbell, Julian. "A Rape in Cyberspace." *Village Voice.* 23 December 1993. Accessed 2 August 2010. http://www.juliandibbell.com/.

Engeler, Amy. "The ID Theft You Haven't Heard of . . . Yet." *Good Housekeeping.* Accessed 15 August 2007. http://www.goodhousekeeping.com/.

Gibson, William. *Neuromancer.* New York: Ace Books, 1984.

Grosz, Elizabeth. *Volatile Bodies: Toward a Corporeal Feminism.* Bloomington: Indiana University Press, 1994.

Hansen, Mark B. N. *Bodies in Code: Interfaces with Digital Media.* London: Routledge, 2006.

Haraway, Donna. *Simians, Cyborgs, and Women: The Reinvention of Nature.* London: Routledge, 1990.

Hayles, N. Katherine. *My Mother Was a Computer: Digital Subjects and Literary Texts.* Chicago: University of Chicago Press, 2005.

Kendall, Lori. "Recontexualizing 'Cyberspace': Methodological Considerations for On-Line Research." In *Doing Internet Research*, edited by Steve Jones, 57–74. Thousand Oaks, Calif.: Sage Publications, 1999.

Kennedy, Helen. "Beyond Anonymity, or Future Directions for Internet Identity Research." *New Media and Society* 8, no. 6 (2006): 859–76.

Locke, Christopher, Doc Searls, and David Weinberger. *The Cluetrain Manifesto*. New York: Basic Books, 2000.

Maupin, Armistead. *The Night Listener*. New York: HarperCollins, 2000.

Rheingold, Howard. *The Virtual Community*. New York: Basic Books, 1993.

Turkle, Sherry. *Life on the Screen: Identity in the Age of the Internet*. New York: Simon and Schuster, 1995.

Viegas, Fernanda B. "Bloggers' Expectations of Privacy and Accountability: An Initial Survey." *Journal of Computer-Mediated Communication* 10, no. 3: article 12.10. April 2005. Accessed August 2007. http://jcmc.indiana.edu/.

Wegenstein, Bernadette. *Getting under the Skin: Body and Media Theory*. Cambridge, Mass.: MIT Press, 2006.

White, Michele. "Too Close to See: Men, Women, and Webcams." *New Media and Society* 5, no. 1 (2003): 7–28.

11   `<?php>`

"Invisible" Code and the Mystique of Web Writing

HELEN J. BURGESS

PHP is a popular scripting language used by hundreds of custom Web applications, widely used software such as WordPress and MediaWiki, and popular Web sites such as Facebook and Digg. Though it was originally designed for personal home page production (hence the acronym), PHP is most often used in server-side scripting. A core part of the LAMP stack (Linux, Apache, MySQL, PHP), it is probably the language most often used for database-driven Web applications, thanks to its easy support for MySQL. PHP code is also popular because it can be mixed freely with HTML, via simple opening and closing `<? php ?>` tags. Even novice users can cut and paste code into Web pages and perform complex database operations—or create security and privacy problems: PHP is involved in a huge percentage of the vulnerabilities tracked by the National Institute for Standards and Technology.

IN AN AGE OF WRITING AND CODE, we are all mystics. In my capacity as a teacher, I am the imparter of mysteries, the encoder of information. My favorite moments have neatly bookended activity in my Web writing, design, and development classes. The first is the moment of revelation, the pulling aside of the curtain: I stand at the podium, enter a URL, and then choose "View Source" in the browser. Students gasp. It's very satisfying.

My second favorite moment occurs at the other end of the class. A student (not a Web designer; perhaps a human development major, or an English student, or a psychology intern) is working on her tags, trying to figure out what is happening on the page. We work together, heads not quite touching, poring over the symbols on the screen, looking for (and finding) the code that will enable the page to display. She utters a small sound, makes a change, and previews the page. And the curtain falls, the page appears—but with a difference. For now, we understand the trick.

In new media studies, we have (or should have) gone beyond simply pointing to a Web page and commenting on its structure, development, and language. But the current trend in focusing on "visual literacy" tends to emphasize what's on the screen, rather than what lies beneath. Our understanding of code has gotten lost under the layers of GUI and WYSI-WYG; graphic design teachers (trained primarily in the visual register) pale when students click on the <> button in Dreamweaver, and the mysteries of markup appear. At the same time, though, teaching plain vanilla markup is somewhat old-fashioned. Sure, there is the magic moment of revelation, but in an age of database-driven pages and invisible scripting languages (PHP, Perl, ASP.NET, the usual laundry list of server-side applications), it seems rather quaint to be teaching the `<p>` tag (or even worse, the deprecated `<font>` and `<table>` tags). Server-side scripting languages complicate markup by writing code for us; if we try to view the source, all we see are the traces left behind in the shape of inert HTML tags. The magic happens elsewhere. Markup, then, has come full circle: from the mystery of the Web page, to the revelatory moment of "View Source," to the invisible scripting we know is there but can't quite get to. Unable to scan the page and take an educated guess at what's going on under the hood, students face a key disadvantage as they try to understand how the Web works. This is especially true in the world of the social Web, which is dependent on database calls and the run-time restructuring of pages for its vitality. "View Source" is no longer sufficient as a mechanistic view of the way information appears on the screen.

In this essay I want to talk about the history of invisible code and its relationship to the mystique of writing. N. Katherine Hayles notes that code has a tendency to operate through "practices of concealing and revealing" (54), in which the programmer chooses which code to leave visible for the purposes of authoring and debugging. But the process of revelation and concealment in markup goes much further back than electronic texts—in fact, it is an essential part of the hidden history of print and writing itself. Long before the magical moment of "View Source," print and book producers were already using their own forms of hidden markup: the symbols written on texts that contained instructions or marked points for the purposes of textual reproduction. These printers' marks are the antecedents of today's markup schema: they are marking up manuscripts in the same way we mark up electronic texts. Therefore, I want to start out by looking at the origin of printers' marks in the era of the manuscript, with the understanding that these marks represent an ancestry of sorts for Web writing.

The history of the marks that constitute any text has long been a field of study. D. F. McKenzie, in his study of the history of bibliography, notes that the body of the text has long held primacy over the other material that accretes to it. He argues for a "sociology of the text," suggesting that bibliography, marginalized (sometimes quite literally) to a supporting role in the text, actually has its own complex social history (15). Jerome McGann, similarly, has made arguments for the crucial and constitutive role of "bibliographic codes" (57): ways in which the material construction of the book (using leading, typeface, layout, gutter, and printers' marks) fundamentally changes the nature and meaning of any given text. Each text thus generates its own history as it passes through multiple marked-up printers' editions.

Markup works similarly in the formulation of historical (electronic) texts. It has its own history (the versioning of SGML/HTML/XHTML), its own grammatical lineage (the development of some tags over others), its own narrative (the archaeological layers of comments attached to shared code), and even its own politics (language choices, browser compliance, and the choice to share code or retain its mystique as the writing of an invisible professional). Markup thus becomes a kind of ghostly writing dependent on context and history, rather than merely a means of formatting text.

It's my belief that we are going through yet another stage in the history of markup. The second part of my essay will look at the functioning of server-side scripting as another invisible hand in the process of electronic writing. Sometimes practices of hiding and revealing, as Hayles describes, are necessary for the benefit of the programmer and/or reader—she cites on the one hand the practice of collapsing/hiding code into object chunks for ease of use, and on the other hand the revealing of HTML and comments for the explication of the marked-up text (54). The HTML tags we can see in our browser's "View Source" window are akin to early printers' marks: they are not readily apparent, but they can be read if we know where to look, in the process of flipping back and forth between page and source code. But sometimes, the concealing and revealing is the result of the operation of the server itself. Because server-side commands execute before the HTML is written, they are hidden from the browser not by the programmer but by the server itself. PHP, currently one of the most popular server-side scripting languages, serves as a useful focal point for my discussion of hidden code: it is written but never viewed by the end user; it is in itself a kind of writer of code, passing instructions to server and thence to a database; it is never seen but only experienced as the end result on the page. Thus, I believe, we are making another transition in the history of (electronic) texts: from visible markup to invisible code.

Working in the Copy Shop

The prehistory of electronic markup is intimately tied in with the history of copying. Texts and markup have been entwined since at least the monastic period of book production through the systematic processes developed to reliably copy religious (and later secular) texts. Writing, as Walter Ong notes, is a codifying of orality into a text that can be repeated, what Ong calls an "exactly repeatable visual statement" (124–25). Any such repetition will necessarily involve a whole invisible apparatus: the infrastructure necessary to carry out the reproduction of texts, including the hidden labor of copyists, who communicated to each other during the production of each text using an abbreviated and highly specialized series of codes. These codes were the language that ensured a new kind of faith—not in the mysteries of religious life but in the fidelity of the text.

The history of organized production line copying in the West goes back to the scriptorium, a kind of workshop for the copying of texts, in which copyists worked side by side with illustrators and illuminators for several hours every day, painstakingly reproducing religious texts—the texts of mysteries. However, the ramping up of the production and copying of books begins with the growth of medieval universities and the rising demand for scholarly books (Febvre and Martin 20). The growth of the secular market for books soon developed into a rationalized system of production: the pecia system, wherein copy texts, known as exemplars, were loaned out in parts (piecemeal, or "pecia") to be copied and then returned. The pecia system was most important in its capacity as a preserver of the fidelity of the text. The loaning of an exemplar meant the limiting of degrees of freedom from the original—a way of ensuring that errors did not propagate over multiple copies.

The pecia system was important in the way it rationalized and specialized the book production system into a kind of laboring human machine. Febvre and Martin note that the pecia system encouraged separation of skills and a division of labor, such that "it became more and more common for separate workshops to be set up, with copyists in one shop, rubricators perhaps in another, and illuminators in another" (26). Along with specialization comes the problem of communication: how does one preserve the fidelity of the text when it must pass through multiple hands for copying, illustrating, and binding? The pecia system achieved this by the use of pecia marks: marks added by the copyist to communicate where each pecia was to be placed. These codes, like the XHTML markup we use today, were specifically meant for the structuring and formatting of text, marking the beginning and ending of a section, for example. Books were broken into sections and handed out for copying. Because this piecemeal approach often resulted in the copied piece starting and/or finishing in the middle of the sentence, the end of a section was marked by the copyist with a pecia mark. For example, "p4" might mean "end of pecia 4." This operated similarly to the "quire signature," which marked the end of a quire (that is, a section of a book made from four sheets of parchment or vellum folded and then stitched, usually resulting in eight or sixteen pages) and ordered it for binding.

Most importantly, from our perspective, pecia marks were an expression of a new way of thinking about texts: as information to be structured and processed. Febvre and Martin note that the medieval copyist often placed instructions on the manuscript to tell the illuminator what to put in (a lion, a garden, a snake). As a placeholder, a guide letter was put in the space where the illuminator (who often did not read the text) was to illustrate the letter (26). But pecia marks were unique in terms of their function: they did not mark the insertion of content, but the beginnings of an understanding of logical structuring—that is, the assembly of the book as a number of pieces that must be stitched together in physical (not semantic) order. The pecia mark is a coding system for the reassembly of a text: a communication of repeatable formatting from one writer to another. This, I would argue, is the beginning of markup: a language specifically developed for the logical structuring and later formatting of a document.

Going to the Print Shop

The movement from the pecia workshops to the print shop is a complicated tale, even from a mechanical standpoint: the trial-and-failure of metal castings, poor paper manufacture, and ink viscosity. The history of the printing press begins with a secret. In a lawsuit in 1434, Johannes Gutenberg was accused of working with partners on "secret processes" (Febvre and Martin 51), including the possible manufacture of press pieces (probably die-cast molds). He was not the only person working in this area—many metalworkers were attempting to improve the wooden block

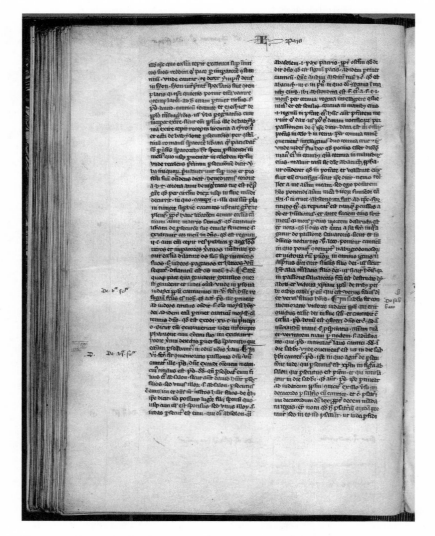

Figure 11.1. Manuscript showing pecia mark in outer margin of verso, "Finis xxx pet." From Guillelmus Durantis, Rationale Divinorum Officiorum. Italy (?), s. XIVin. call no. HM 26298. Reproduced by permission of The Huntington Library, San Marino, California.

printing process by creating a system of interchangeable letters or moveable type. But Gutenberg is significant as a precursor of a dot-com millionaire: he was the holder of the technological "secret of the art of printing, . . . the making up of a page of print from separate, moveable types" (50). Already the value of the secret was evident: Gutenberg's angel investor, Fust, found him difficult to work with, so he waited until he had trained his more pliable assistant, Peter Schoeffer, and then called in the loan. Gutenberg's secret was thus revealed, and he died bankrupt (Febvre and Martin 55).

During this period, another series of codes was developed for marking up the text. The register—a table recording the first and last word on every page—was developed so that the printer could ensure pages were in order without having to read the entire text. Letters and numbers appeared on the bottom of leaves, not for pagination (as a service to readers) but as a way of ensuring the pages were folded and bound in the proper order.

If the pecia system enabled the reproduction of the text to a reasonable standard of fidelity (by using an exemplar), the printing press was able to capitalize on its mechanical ability to fix the page many times over from the same tray. Although errors could propagate this way (if the tray was laid out incorrectly, every single copy would also be incorrect), and in fact made for a corruption of the text that could be more damaging precisely because of its consistency (hence the introduction of printers' editions), the printing press allowed for a fundamental shift in the culture of writing through its emphasis on mechanical reproduction. The printers' marks on a printed copy were evidence not of the faith of the copyist in a religious sense—or even faith in the sense of accurate fidelity to an original manuscript—but rather faith in the ability of the machine to automate the material process of reproduction.

The Archaeology of Electronic Markup

The movement from religious to secular to mechanical faith has taken another turn with the implementation of document processing for online environments. Although computerized print software is essentially an analog of workshop procedures and relies to a great extent on precisely the same printers' conventions and vocabulary, HTML, descended from the more general schema of SGML, is designed as a way to facilitate the formatting of electronic information in a logical fashion explicitly for display online, in a browser. Like the regulated procedures of the scriptorium,

HTML is designed with reproducibility in mind: the faithful rendering (in this case, electronic display) of the same file over and over. HTML tags are, in this case, similar to the guide letters and notations written into the manuscript: instructions to the browser for formatting the file on the screen. But at the same time, the file is not expected to be literally copied: no second file is produced. This is a move to a kind of virtual faith—a faith in the consistent display of a document that will nevertheless disappear when the window is closed.

Electronic markup is a fairly straightforward, mostly human-readable system in which each piece of content is tagged. For example, I am working up a simple Web page in HTML to display classroom grades. Here's what my code looks like:

```
<table>
<tr> <td>James</td> <td>Jones</td> <td>B</td> </tr>
<tr> <td>Pavithra</td> <td>Jones</td> <td>B</td> </tr>
</table>
```

Tables such as the one above are an interesting case study in the way HTML has changed over time. Originally, HTML tags were designed wholly for logical data display: the heading <h1> is larger than <h2> to signify relative importance and nesting, not to give the letters a nice fat 16-point Times New Roman look. Tables, in their original formulation, were designed to be just that: a method of tabulating data accessibly. As higher bandwidth speeds and more efficient image compression algorithms enabled the routine use of images, semantic markup started to get tangled up with visual formatting. The graphic designer's desire for flashy pages and fixed-size layout complicated the split between content and logical structure by using structural elements (notably tables) to visually format pages. Ironically, this visual formatting relied on invisible markup and image elements. Rather than being used as a logical grid to organize data, HTML tables were used as a grid for laying out sliced-up images intended to produce pixel-perfect visual layouts. The one-pixel transparent GIF (an invisible square) was routinely used to force page layouts.

Groups such as the World Wide Web Consortium (W3C), dedicated to the formalizing of Web standards, and a few designers (notably Jeffrey Zeldman and Eric Meyer), who wanted code to be more compliant with browsers and easier to reformat, were unimpressed by the hybridization of logical structure and visual design. Their dissatisfaction led to the standardizing of XHTML and CSS, a rigorous approach aimed at again separating form and content. Such a form/content division is logical and readable: "View Source" once again becomes a human HTML reader's best friend. We can read the markup and style sheet and make a fairly accurate guess as to what the document would look like on the page. And yet even at this stage, we already see the beginnings of a deferral of content: the link to an external style sheet means we must find and read that file to get a sense of the look and feel of the page. Nevertheless, the style sheet is readable by the end user; all the information is there. Reading and interpreting markup is thus relatively simple while we are talking about an electronic document in terms of one XHTML "text" to which is applied a structural markup and a presentational markup.

Three or four years ago, I would have been happy with my online grade listing. But mere display is really not all that useful. I've decided I want to put all my grades into a database and rerender the page in a scripting language, so that I can make changes to grades and add students without having to rewrite all that code. Virtual fidelity is no longer my goal: I want to change the document from reload to reload. In this instance, HTML is inadequate. I need a place I can store my grades, a way to update them, and a way to display them through my browser (computer scientists call this CRUD—create, read, update, delete). In short, I need a scripting language that can interact with a database.

PHP 101: What Is It? How Does It Work? And Why Should I Care?

PHP (in a typical piece of GNU-recursive jokery, this stands for "PHP hypertext preprocessor") is a server-side scripting language that, among other things, can be made to write code. In HTML, we are accustomed to writing the following:

```
<p>Pavithra Jones</p>
```

The browser reads these instructions, and behaves accordingly, printing to the browser

Pavithra Jones

In PHP, however, we have another kind of operation going on. Usually, you upload your HTML files to a server. Browsers all over the Web send a request for the file. The server gives it to them, and they read the HTML and display it accordingly.

PHP adds in an extra layer:

- The browser asks for the PHP file.
- The server looks for the file, executes the PHP, and gives the result to the browser as HTML.
- The browser interprets and displays the HTML.
- We see the words "Pavithra Jones."

Think of a restaurant: When we go to the restaurant and ask the waiter for an egg sandwich, he doesn't go back to the kitchen and dig up an egg and some bread. He goes back to the kitchen, asks the chef to cook the egg and toast, and then brings them back to us prepared. In this case, the Web server is the chef, cooking up the PHP code and sending it back as HTML.

The PHP code, in fact, looks like this:

```
<?php echo "Pavithra Jones"; ?>
```

Again, we will see

Pavithra Jones

But if we look at the HTML source, the PHP will have magically disappeared, leaving plain vanilla HTML identical to our first example. The difference is that the server is reading and executing the PHP portions of the code, and writing "Pavithra Jones" when it has been told, right into that dynamically scripted Web page.

This simple echo command might be interesting in that it represents the movement from simple tagging to server interaction, but thus far it merely replicates the functionality of the HTML page: to display a hard-coded message. Why bother, when hard-coded HTML achieves the same purpose, and without having to compile PHP on your server? What gets interesting is the magic that happens when PHP interacts with a database language such as MySQL to retrieve information and display it on the page. I've decided I only want to display grades for students with the surname Jones. Let's take a sample snippet:

```
<?php
$result = mysql_query("SELECT * FROM DTC355 WHERE sur-
name='Jones'");
?>
```

This is a hybrid piece of code. The **SELECT** part inside the quotes is a MySQL query: a piece of code asking the MySQL server to select all the records in the database table named "DTC355" with a "surname" value equal to the string "Jones." `mysql_query()` tells the server to execute the query inside the quotes. And the record data gets fed into a variable called `$result`. From there, it's a simple matter to get it to print on the page:

```php
<?php
// Get records from the "DTC355" table
$result = mysql_query("SELECT * FROM DTC355 WHERE surname ='Jones'");
// keeps getting the next row from the database until there are
no more to get
while($row = mysql_fetch_array( $result )) {
// Print out the contents of each row and concatenate into lines
echo $row['name'] . $row['surname'] . $row['grade'];
}
?>
```

All the information contained in `$result` gets fed into an array called `$row`. Then `$row` prints its values over until there are no more records in the array. This will produce the output:

JamesJonesBPavithraJonesBSammyJonesCChandraJonesA

Clearly we have all the information, but we want it to be marked up usefully. This is where PHP comes into its own as a metamarkup tool:

```php
<?php
// Set up some repeating HTML tags to save some space later on
$startrow="<tr> <td>";
$cell="</td> <td>";
$endrow=</td> </tr>";
// start the table tag in HTML and put in a new line string feed
echo "<table>" . "\n";
// Get records from the "engl499" table
$result = mysql_query("SELECT * FROM engl499 WHERE surname='Jones'");
// keeps getting the next row until there are no more to get
while($row = mysql_fetch_array( $result )) {
// Print out the contents of each row and concatenate with HTML
tags
// use the new line string feed to break up the HTML when viewing
source
```

```
echo $startrow . $row['name'] . $cell . $row['surname'] . $cell .
$row['grade'] . $endrow . "\n";
}
// close the table HTML tag
echo "</table>";
?>
```

The PHP script has now taken the information from the database, added in HTML for formatting, and output it. We've added in some invisible new line characters (\n) so that when we view the HTML, it won't all be in one long string. When we view the source, we see this:

```
<table>
<tr> <td>James</td> <td>Jones</td> <td>B</td> </tr>
<tr> <td>Pavithra</td> <td>Jones</td> <td>B</td> </tr>
<tr> <td>Chandra</td> <td>Jones</td> <td>A</td> </tr>
</table>
```

Not pretty, but useful as an example: the PHP scripting has disappeared, to be replaced with a machine-written, static page.

This kind of query-based scripting, combined with PHP-generated markup, is interesting on several accounts. The first is that it is still displayed flat: all we see if we view the source from the browser is the HTML to display it. The second is that we can make as many calls as we want into the database: the page is suddenly as malleable as the database itself, a plastic space that can change with every impatient reload. The third is that it's writing within writing within writing: HTML holds the PHP, which holds the MySQL query.

But the problem with PHP is that the end user can't read it. Now we are talking about a disappearing or invisible tagging system—one in which not only is the PHP operating invisibly on the page, but also the actual code of the document can no longer be read in the "View Source" command. Because PHP is executed on the server, not in the browser, any attempt to view the source locally reveals only the executed markup, not the PHP commands. We have moved to a tripartite markup system: content included (via script or database call) by PHP at run time, resulting information marked up logically in the XHTML, final look-and-feel text/image formatting in the CSS. This movement, from presentation to dynamic generation and retrieval, echoes the medieval transition from religious manuscript production to secular copy production: from the guidelines and sketches of the monastic scriptoria to the operational marks of the pecia workshop, and later the page registers of the print shop. An emphasis on the visual formatting of an existing long-form document gives way to instructions for the logical structuring of a document in pieces: the document object model. The shift from static formatting to run-time assembly also echoes the pecia and print shop's rationalization of production, in that pieces are farmed out to different agents—the style sheet, the looping algorithm, the database—just as pieces of a book were given to typesetter, binder, and cutter. To the end user, all this labor is invisible.

Ghosts in the Deep Web: The Database

In an electronic file, PHP scripts stand in for the information they will call; like pecia marks, they pinpoint the beginnings and endings for inclusions of text, whether quire, chapter, or blog entry. Like printer's marks, PHP snippets elicit an operation or execution: they are symbols requiring an action be performed. The key difference is that although pecia marks are meant to ensure repetition, PHP scripts are written to ensure flexibility; the goal of a dynamic Web site is to produce pages customized to each query. PHP does not contain the information itself (this is left to the database), but rather contains instructions on what to do with the data, and where to put it.

Database storage and retrieval (known as the deep Web) is obviously useful for corporate databases, shopping carts, online ordering, online catalogs, and so on. But it's interesting to us as Web writers because of the next-generation social Web. PHP scripting allows us to include a chunk of (often unformatted) content (notably, content presented by a content management system [CMS], blog entries, menu items) from a database, place it on the page, and then apply markup and styles. It allows for the switching of styles and templates according to whim, the easy addition of information via a text box and the reshuffling of that information or narrative at a moment's notice. A blog or a CMS can be changed with a click from a chronological sorting of entries, to one by category, to one by author.

The flexible chunking system of the deep Web means several things for a writer. First, content is broken into discrete sections and stored in a database, which has the effect of breaking the historical tradition of writing as a kind of complex narrative weaving. Lev Manovich has identified what he sees as a fundamental antagonism between narrative and database. Database logic, he argues, stores information in discrete pieces that are only connected together at run time: the database "exists materially" (in the sense that all the available data are stored somewhere ready to be used) (231), while "the narrative is constructed by linking elements of this database in a particular order, that is by designing a trajectory leading from one element to another" and is thus "virtual" (231). Fundamentally, a database is about storage, while narrative is about ordering.

Narrative personal blogs, threaded discussion boards, and news services rely on just such a logic. On the surface, such sites appear to work chronologically (usually, although not always, backwards, with the newest entry at the top). Items might be linked in a kind of mininarrative. For example, Boingboing.net often posts a minichronology at the bottom of a post, pointing to recent posts on the same topic, which are also listed chronologically. But in fact, the flexibility of PHP to retrieve multiple records from a database and display them according to any number of requirements means that at any particular moment, the content might be reordered on the page by date, or category, or subject—or, in the case of multiauthored sites, the author. Under the hood, we don't have to move information from one category to another. We just change category flags in the database, or add a new keyword for multiple categories. Information retrieval can now be keyed to specific tags and sorted according to a different chronology, author, or numerical pattern. The SQL command "**ORDER by [asc, desc]**" renders this operation trivial. The server can take multiple database entries, recombine them into different groupings, and display them out of sequence, thereby having the effect of changing the fundamental nature of the page as progressive narrative. Narrative time (in terms of the chronology of the entry) is overturned in favor of recombinant time.

This reordering property fundamental to database logic is also spatial, as evidenced visually in the GUI to a MySQL database. One example, the Web-based phpMyAdmin panel, displays data spatially in text boxes and tables as a way of echoing its internal structure. This is only a visual representation, of course, but it resembles the spaces on an annotated manuscript page. The difference is that the spaces of the page can be changed at run time; clicking on a table heading or running a query will produce a new screen with a different combination of tables and text boxes. Unlike the old manuscript pecia marks, executed with repeatability in mind (the faithful reproduction of a text), PHP queries produce texts that can differ according to changeable input (and the data itself). Attempts to "browse" a database at the back end with a GUI such as phpMyAdmin lead to a misrecognition. We may think we are looking under the hood at the database,

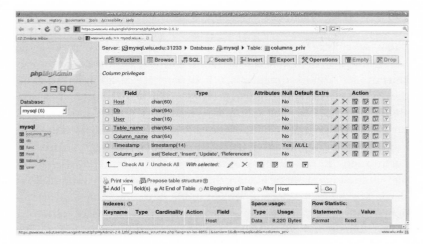

Figure 11.2. Tablelike layout of phpMyAdmin MySQL administration software.

but at any single moment, we are looking at a visual and tabular representation of a much more abstract field of data. The design of the interface itself determines what we see of the data structures and information.

At the same time, the phpMyAdmin interface offers us an interesting example of the meta-level similarity and difference between pecia mark and PHP query. Clicking on a link in phpMyAdmin displays the resulting data, but also helpfully displays the underlying query (the actual coded request) that was performed. On a manuscript, the pecia mark is already there, waiting to be executed. In the phpMyAdmin graphical interface, time is flipped: the query is performed, and then the code is displayed. This is because, ironically, the interface is itself scripted in PHP.

A second consequence of the PHP-enabled "chunking" of writing is that a site with a wide and changing range of dynamic content becomes one page; the database is the storage area for content. At its simplest, a site can now consist essentially of one XHTML page containing PHP scripts, one CSS page, and a database. Even CMSs, which seem hugely complicated in their file structure, adhere to the basic rule: the page you see is always the same index.php page. It calls various scripts and includes smaller formatted chunks (for example, a sidebar template or a footer) from other files, but the database is still responsible for all the content. At run time, the server executes the scripts in order, gathering first the "include" templates such as the sidebar, and then slotting in data from the database.

Within that simple structure, however, one must make a series of very careful choices. The retrieval of information is flexible, in the sense that databases change over time and can be called in multiple ways, yet at the same time rigid, in that the author has to decide in advance what choices can be made. This restriction affects both database design and PHP script design: what is likely to be most useful as a cluster of information? How should one piece of information be related to another in the database? My classroom example is simple enough—all I am likely to need to know is a link between name and grade. But complex databases can end up connecting one-to-many or many-to-many pieces of information. For example, my students might be in more than one class, or I might want to share the database with a colleague. I might suddenly discover that I have a student with a name that is too long for the field, or I might discover that one of my students has changed her name to Moon Unit, in which case I'll have to create a new record and somehow link it to the old one.

Holding all this structural information in my head long enough to determine what questions will be asked of it means conceptualizing the whole, even though I will only ever see it in parts. The database itself becomes, in some ways, a kind of material ghost that lurks at the back of every blog and CMS. As Manovich says, all the content (the paradigm) is there, but only a small amount will be syntagmatically displayed at any given page reload.

Finally, the intervention of PHP in the Web writing process means that agency is transferred from browser to server. If we think about it at all, it is the browser that is usually ascribed the agency in a Web session. The browser is, after all, responsible for sending and receiving coded messages; for interpreting the pecia marks embedded in the document. The browser is our faithful copyist, laboring within a series of markup rules to produce a text that is, depending on the browser type and settings, more or less the way the author imagined it.

With a PHP/SQL-driven site, however, the agency moves elsewhere. The secret technology is the server, which has now become more than a server: it does not just parcel out files, but also parses instructions inside them. The server is no longer a server but what Bruno Latour would call an actant—a semiautonomous personality or "quasi-object" (51) that is built in the nexus between different technologies of matter, machine, writing, and consciousness. Along with the end user author entering data and the originating author of the code, we have another hidden operator at work: the server, which faithfully executes the PHP code and writes the results to browser-readable HTML.

In the age of the social Web—notably the advent of mass user-generated content sites such as YouTube—dynamic authoring has taken on a newly decentered role. Indeed, it is no longer clear who—if anyone—can claim to be the author of a site. Writers and video uploaders interface with databases and scripts in a network of relations, with agency shifting between writer, server, database, and browser. Manovich notes that the database is not merely the container upon which the algorithm operates:

> It may appear at first sight that data is passive and algorithms active—another example of the passive-active binary so loved by human cultures. A program reads in data, executes an algorithm, and writes out new data. . . . However, the passive/active distinction is not quite accurate because data does not just exist—it has to be generated. (224)

For the casual user, the emphasis on content often serves to obscure the huge investment that has gone on behind the scenes to produce the infrastructure allowing all these mash-ups and content-sharing sites to exist. Blogging software, wiki packages, and CMSs are painstakingly scripted and debugged by large communities; the customizable Ajax interfaces are the product of intensive start-up development. Open source development means code is more accessible, but much of this code is aimed at producing seamless front ends for data input. At the click of "Upload" or "Refresh" or "Save Category," a kind of discretized and recombinant magic happens, with the server secretly working behind the scenes to produce the prestige. This last is the most interesting because it leads us right back to the beginning of writing: writing as mystique, the domain of the invisible expert who executes the marked-up page.

Conclusion

Server-side scripting of database calls and markup are bringing into obsolescence the end-user "View Source" command. To be able to read HTML is no longer sufficient—one must be able to read the logic of the page and be able to tell what is going on underneath without being able to read it directly. Similarly, back-engineering a dynamically driven site is becoming a necessary skill for Web authors as they participate in the community of Web development. But marking up is not the "translation of the world into a problem of coding," as Donna Haraway would say (164). It is the act of making visible the invisible.

With this injunction, we have two options as readers and reproducers of electronic texts. The first, if we are unlucky, is to have to work through inference and perform the indirect act of reading the screen: looking at logical divisions of content in the page and working out the probable database calls. This requires a careful training in not merely visual (in the sense of images) but spatial and quantitative literacy—an understanding of how databases are organized, how they can be queried, and where and how the data can be best displayed on the page. In addition to the database design, the architecture of the site itself can often require a complex act of inference. Attempting to read a site can result in a series of structural deferrals as we look for a piece of code. In a CMS such as Joomla, for example, a desire to change the visible front page index.php will lead us several levels down the directory structure to another index.php inside the "Themes" folder, hacks of .htaccess files redirect pages while hiding the actual directory structure in the URL; server-side includes import of many snippets of HTML and PHP from other parts of the site. Thus, in addition to database logic, we must learn to read site design logic.

Our second option, if we are lucky, is to learn to read the code itself. In this endeavor, we have a notable helper: the comment. Comments are the pecia marks, registers, and guide letters of our time—in many ways, the literary exegesis of code. As creators of digital texts, we learn early on (if we are taught well) that commenting code well is absolutely essential for the transmission of an electronic text from one person to another, whether they be a fellow student, a fellow designer, or a client. The comment is in itself a literary form. The pinnacle in the genre of commenting is, ironically enough, a print text: *Lions' Commentary on Unix* (often called the *Lions' Book*). This book is a listing of the entire source code of the Unix 6 kernel, with an accompanying commentary. It bears a distinct resemblance to medieval text commentaries.

The comment is particularly vital today in the modern equivalent of the scriptoria: the large, virtual workshops of the open source software development community. Open source tasks in such communities as Source-Forge are parceled out in pieces for coding by individuals; small software tools and plug-ins are developed by multiple authors using carefully arranged CMSs (such as Subversion and Bazaar) for versioning and communication. Chief among these communication tools are the comments embedded within the code itself: section markers and small reminders from one developer to another, explaining what a particular piece of code does or what interactions it might potentially have with other pieces of software.

Comments are both like and unlike the original pecia marks they resemble. Structurally, they do indeed mark out the beginnings and endings of sections, and they mark out spaces for the inclusion of data. But just as PHP has changed the way we view markup—by becoming an active author in the process of generating a site from a database—so too comments go beyond the logical markers of the pecia system. Instead, comments seek to explain what will happen when the code executes; they look into the future and tell us what the invisible machine will do with all that data. Comments are thus a crucial companion to executable code, just as HTML was a companion to simple text and pecia marks were to manuscript fragments.

As I have shown, the social history of texts—a history of faith, from religious piety, to secular fidelity, to mechanical reproduction, to electronic display—that began in the medieval workshop has undergone a clear transition again in the age of the database and server-side script. But through all of these transitions, we have seen the consistent use of textual marks: guideline, pecia, tag, script, comment. In an electronic culture where so much press is devoted to front-end social tagging, user content development, and online jabber, we forget that there is always another kind of communication going on, an invisible social code: the communication between one developer and another, one server and another, one database and another. These marks may be hidden, but they are the underpinning of electronic writing.

Works Cited

Febvre, Lucien, and Henri-Jean Martin. *The Coming of the Book: The Impact of Printing, 1450–1800.* Translated by David Gerard. 1976. London: Verso, 1997.

Haraway, Donna. *Simians, Cyborgs, and Women: The Reinvention of Nature.* New York: Routledge, 1991.

Hayles, N. Katherine. *My Mother Was a Computer: Digital Subjects and Literary Texts.* Chicago: Chicago University Press, 2005.

Latour, Bruno. *We Have Never Been Modern.* Translated by Catherine Porter. Cambridge, Mass.: Harvard University Press, 1993.

Lions, John. *Lions' Commentary on UNIX 6th Edition with Source Code.* 1976. San Jose, Calif.: Peer-to-Peer Communications, 1996.

Manovich, Lev. *The Language of New Media.* Cambridge, Mass.: MIT Press, 2001.

McGann, Jerome. *The Textual Condition.* Princeton, N.J.: Princeton University Press, 1991.

McKenzie, D. F. *Bibliography and the Sociology of Texts.* 1986. Cambridge: Cambridge University Press, 1999.

Ong, Walter J. *Orality and Literacy: The Technologizing of the Word.* 1982. London: Routledge, 2002.

12 FROM CYBERSPACES TO CYBERPLACES

<img>, Narrative, and the Psychology of Place

RUDY MCDANIEL AND SAE LYNNE SCHATZ

<img> *is an HTML and XHTML element used to embed images in a page. Writers can control its size via* height *and* width *attributes and can specify alternative text via* alt *attributes. The original HTML specification did not contain an* <img> *element; it was proposed by Marc Andreessen in February 1993 on the WWW-Talk mailing list. Andreessen's Mosaic browser is often credited as being the first browser to display embedded images. Web writers adopted the tag quickly, and* <img> *soon became ubiquitous, fueling the need for image hosting and sharing sites such as Ofoto and Shutterfly. Today, Flickr, one of the most popular of such sites, extends social networking to images: Flickr users share, tag, and apply each other's images either through direct embedding via* <img> *or through more extended embedded code (where* height, width, *and* alt *are predefined by Flickr). By referencing a file outside the Web document—and quite possibly on another site—*<img> *activates Ted Nelson's theory of transclusion, the usage of external elements by reference, not by inclusion.*

THE WEB CAN BE CONCEPTUALIZED as an aggregation of disparate sites; as Lev Manovich suggests, virtual spaces "are most often not true spaces, but collections of separate objects" (253). However, in this chapter, we propose that the Web is not simply an assemblage of data, but rather that it is a collection of active places that individuals construct in their minds. We argue that without imagery (that is, without the <img> tagging capabilities) the Web would be less able to facilitate rich, recognizable, and communal narratives. Without pictures, icons, photographs, and other graphical representations, it would be conceptualized as more tool-bound than experiential, more of a sterile informational resource than a collection of organic places. Yet to move from Manovich's "collections of separate objects" to our notion of a shared community, two conceptual leaps must be made. We must first connect hypertext documents in a method sufficient to create a shared space, then imbue this space with an atmosphere of personality—which can be expressed through causal and temporal relations—that shifts that space into a more meaningful and memorable place. Through the powerful <img> tag and its associated textual attributes (such as alt and longdesc), this is possible by means of visual storytelling techniques. In this chapter, we will discuss the notion of place as a psychological construct and explore online imagery from a rhetorical and narratological perspective. We also discuss some techniques from environmental psychology that are useful in understanding imagery as a cognitive device that aids in narrative comprehension.

We first argue that the unique cognitive situation afforded by online imagery enables people to better orient themselves in virtual space and make sense of spatial information thorough the construction of schematic narratives. Next, we posit that the rhetorical power of graphical representations allows individuals to better ascribe shared meaning to their online experiences. In this sense, Web sites exist as communal narratives or narrative fragments, potential units waiting to be assembled into available stories about everything from product purchases and customer service to personal anecdotes or emotionally charged and cathartic outpourings of sentiment. We will use literature from psychology and narrative theory to discuss the ways that the <img> tag and its attributes were—and still are—integral to shaping the narratologically rich Web of today. Specifically, we will apply concepts and techniques such as legibility, image-ability, wayfinding, and image schemata to our rhetorical analysis of online imagery as an intersubjective and intertextual facilitator of narrative place.

Precursors, History, and Theoretical Basis

Precursors

For many contemporary Web users, it is difficult to imagine the World Wide Web without imagery. Even today's most rudimentary Web sites rely heavily on imagery; commercial sites feature more sophisticated themes and amalgamations of still images and, increasingly, animations and even high-quality video. The everyday Web user is likely to encounter a great variety of visual information—from the simple colored backgrounds and stylized text used in blogs such as WordPress to the complex graphical layering of Google Maps and the robust photographic manipulation and tagging tools of Flickr. There was a time in the not-so-distant past, however, when pictorial and graphical representations were nowhere to be found.

In his analysis of early networked experiences, Mark Nunes explains how the evolution of textual Internet technologies led us to conceptualize virtual space as a location in which activities take place. In *Cyberspaces of Everyday Life*, he discusses the emergence of telepresence, or the idea of projecting oneself within a virtual medium and replacing real world cues with virtual ones, and its interesting implications for spatiality during his early experiences with MUDs (multiuser dungeons) and MOOS (MUDs, object oriented). In these environments, players would meet to have conversations and engage in virtual interactions, and yet the events "literally take place neither *here* (at the computer screen) nor *there* (at some other location), but rather within the medium itself" (3). Nunes then considers the implications of the graphical user interfaces of Web browsers as particular mode of mapping and navigating virtual space. He notes how the abilities of the first Web browsers allowed us to conceptualize Internet documents as "areas" of exploration linked both to previously visited areas and various permutations of potential "next" areas (3).

Adding imagery enriches the narrative quality of online activities by simplifying two activities: the production and consumption of personalized content. This is particularly true in regard to larger communities of visitors and the narrative history that they provide to a site. Images can facilitate these processes in several ways. First, consider individual image postings, a simple example of which is found in the use of personalized avatars. When posting a link or thread to a discussion forum, visitors can easily configure a unique avatar to represent an element of their personality or identity. Commonly, visitors will choose to select images of movie stars, pets, astrological signs, personal objects, or even favorite automobiles to embed an idiosyncratic element along with their textual contributions to a site. By doing so, they may combine the fantastic with the ordinary, the personal with the impersonal, or, at the very least, the self with the community. We argue that they do so in a way that is more immediate and visceral to audiences than what would traditionally be expressed through a text-only signature. Such avatar images can also be hyperlinked back to personal spaces, thereby claiming and linking a small chunk of virtual real estate as though one were planting a flag on foreign soil or leaving a business card for future negotiations. This avatar then becomes a narrative fragment, a potential protagonist waiting to be used in a variety of online stories. Or perhaps the avatar image is waiting to play a key role in a larger metanarrative sustained by an online community.

Another way to facilitate shared narrative is through the coordinated use of many images that together build a comprehensible Web site design. Today, even the most barren sites may use images to define discrete sections (such as a special background for the navigation), highlight key objects (such as the use of icons for important links), and add symbolic meanings to sites (such as a specialty font applied to a page title). Most sites comprise numerous images, and the summation of these creates a unified gestalt that infuses sites with narrative power and emotional meaningfulness.

Theoretical Base: Image and Text

From the perspective of narrative, what is interesting about the `<img>` tag and its attributes is the degree to which they can collectively produce hybrid configurations of multimodal stories. These stories are composed of both image and text, with the textual content often being minimized in profile by the image. This has unique consequences for the narrative comprehension of online information because what is not explicitly shown or told to the reader is often just as important as what is. Images alone can reveal causality or temporality through sequencing and arrangement, but additional connections can be instilled using textual attributes. For example, Roland Barthes notes how captions added to press photographs can connote or "quicken" photographic images with "one or more second-order signifieds" ("Photographic" 25) in order to degrade or subvert the original narrative of an image. His example is a photograph of Queen Elizabeth II and Prince Philip leaving a plane, with the caption, "They were near to death, their faces prove it," even though at the time the photograph was taken, neither knew anything of the accident they had just escaped (27). Returning to our previous avatar example, similar subversive techniques might easily degrade the psychological value of a place by annotating certain individuals' avatars with misleading labels (e.g., troll). Or, conversely, they might add truth or legitimacy to personal narratives in a variety of other ways through accompanying comments or ratings. Text can serve to subvert or legitimize the narratives created by online images.

Text and image have always had a somewhat rocky relationship when used in hybrid compositions; this dynamic is well documented by theorist W. J. T. Mitchell in his book *Picture Theory*. Here Mitchell explains that image and text have long been used together for a variety of narrative purposes in comic books, newspapers, illustrated manuscripts, and artistic works such as those produced by poet-painter William Blake (89). In comparative media studies, it has become fashionable to analyze one medium through the lens of another, to somehow have a glimpse into the sociohistorical context of artistic production during a particular time or within a particular community. Mitchell cautions, however, that theoretical attempts to compare image and text (such as using literary conventions to study a painting produced in the same period) can lead to misleading (at best) or even pointless results. He writes:

> The most important lesson one learns from composite works like Blake's (or from mixed vernacular arts like comic strips, illustrated newspapers, and illuminated manuscripts) is that *comparison itself is not a necessary procedure in the study of image-text relations.* The necessary subject matter is, rather, the whole ensemble of *relations* between media, and relations can be many other things besides similarity, resemblance, and analogy. (89)

Following this line of reasoning, then, there are various ways of referencing combinations of image and text such as those found in the <img> tag and its textual attributes. To distinguish between various image–text relations, Mitchell follows three different conventions when writing of the relationships between text and image. He uses the term *imagetext* to simply describe composite works combining imagery and text. *Image/text* is used to identify the problematic "gap, cleavage, or rupture in representation" (89) that occurs when trying to compare visual and verbal media using a single theoretical framework. Finally, *image-text* describes the active relationship between text and imagery in hybrid compositions. Following this use of the hyphenated term *image-text* to describe the relations that occur between visual and verbal modes of expression, we can similarly use <img>-text to describe the active relations made possible through the arrangements, juxtapositions, and active transformations made possible when visual and verbal anchors are used together in hypertext documents. In fact, the <img> tag can produce <img>texts, <img>/texts, and <img>-texts.

For example, consider a simple JPEG image of a red rose in bloom. This rose could serve a variety of narrative purposes online. It might serve as a focal point for browsing through potential romantic partners on a dating site, as an overlaid embellishment to provide additional ethos to a particular avatar, or as a hyperlinked sign to a gathering-point discussion board for soliciting gardening stories contributed by flora enthusiasts. Each of these uses is separate in a narrative sense; one contributes to narrative plot, one to character, and one to place.

When encoded for display on the Web, an image and its associated alt attribute might look something like this: `<img src="flower.jpg" alt="A beautiful red rose" />`. This hypertextual code, when rendered in a browser, can be thought of as an `<img>`text, which would emerge when the reader sees the image of a flower while simultaneously seeing the alt attribute's text displayed as she hovers her mouse over the image. To disrupt this narrative pairing and view the components as `<img>`/text, one needs only to recognize how tenuous the relationship is between the alt attribute's text and the linked file name of the image. It is easy enough to change the text or the image and disrupt the harmony of this relationship altogether, perhaps by replacing a healthy and blossoming rose with a dried and withered rose for poetic effect. Suddenly, love is doomed, character is diminished, and a place to discuss gardening tips becomes a place of mourning for what once was. Furthermore, the two types of media produce two very different cognitive outcomes: when we see typed words describing a beautiful red rose, we generally fill in the details of that rose using our memories or imagination. However, when the rose itself is displayed on screen for us to view, we have no such opportunity for recall or creativity. Instead, we are asked to quickly and immediately focus on a particular signifier for that signified concept, not necessarily the one we would produce on our own. It is therefore unreasonable to try and engage with each type of media using the same expectations and rules for narrative production. The rose as image and rose as text are in alignment, but it is easy to disrupt this alignment in a variety of ways, both cognitive and technological. Similarly, the `<img>`-text relationship can be observed through the fleeting nature of both image and text; move the mouse away from the image and the alt text disappears. Or include additional still images as frames in an animated GIF. Rotation, transformation, replacement, and substitution are all active relations that are within the designer's reach. These techniques can be leveraged for various narrative purposes through the use of other linguistic forms such as irony, metaphor, metonymy, juxtaposition, or parallelism.

The fluid and organic nature of hypertext itself also presents several opportunities for thinking about `<img>`-text relations. For one, we can recognize that text can be transformed into imagery through the use of inline tags such as `<div>` and `<span>` and the use of creative formatting techniques involving the scaling, coloring, and font-transformation operations available with Cascading Style Sheets (CSS). Although it does not use the <img> tag for the transformation of its <img>-texts, the CSS Zen Garden is an excellent example of various techniques for translating textual content into imagelike arrangements. Second, and more pertinent to our discussion of the <img>-text, we can consider how its attributes can provide an active dimension to the display of images online. The alt and longdesc attributes provide the ability to annotate imagery with transient captions; as our example illustrates, these texts only appear when a reader moves her cursor over the image, not discriminating between reflective pauses and serendipitous accidents. Similarly, the id and class attributes can be used to show or hide images conditionally by means of emerging development techniques such as Ajax to toggle the visibility of images according to the behaviors of users (Garrett). At an even more primitive level, the browser itself provides potential activities for <img> elements; Internet browsers can be customized; images can be enlarged, reloaded, shrunk, or removed altogether; entire pages can be highlighted, added to show automatically upon the browser's loading, or ignored altogether (added to a blacklist or security filter). Mark Nunes claims that understanding the active relations of computer-mediated communications can move us toward an understanding of cyberspace as a communicative event situated in not only discourse and language, but also performance (12). This is critical in the establishment of narrative place as the reader must play a role in the process of both reading from and writing to online communities.

In the remainder of this chapter, we will explore the discursive performance of images and <img>-texts and discuss how they enable a certain type of narrative—the narrative of place—to emerge from collections of hypertext documents. As we have stated previously, we believe that the idiosyncratic properties of images and <img>-texts on the Web better enable individuals to develop a common shared conception of place in cyberspace. This collective understanding resembles the shared narrative that individuals develop about physical spaces. In both cases, people collectively contribute to a story that implicitly describes the behaviors, values, meaning, and identity associated with a physical—or virtual—location.

Theoretical Base: Narrative

As difficult as it is for many contemporary users of the Web to imagine the early days of networked computing in which textual network protocols dominated online activity and images had not yet infiltrated cyberspace, it is equally difficult to imagine complex forms of communication without storytelling. Narrative allows people to express themselves in a variety of forms and for a variety of purposes; it is the structure upon which complex messages can be woven, so that communication can be easily shared and, in turn, understood by others.

Structurally, stories are fairly easy to comprehend, with a basic set of characteristics—plot, character, and environment—that any individual can relate to on the basis of his or her own life experiences. Narratives are then the expressions of these stories using media. In many cases, stories recycle or repurpose basic and familiar plots that are configurable in an infinite array of dramatically pleasing yet comfortably recognizable variations. Although imagery can most obviously influence the environmental component of storytelling through the placement of scenery, visual depictions of characters, or even the atmospheric use of color or shading, it may also influence character and even plot through the selective placement and arrangement of visual elements.

Barthes tells us that "the narratives of the world are numberless" ("Introduction" 46). With the enormous number of potential stories available to be formed from our raw experiences and observations about the world, it is not surprising that storytelling has become such an integral part of human existence. Through well-established narratives that use techniques such as metaphor and other forms of figurative language, story readers are able to develop shared understandings of character, time, motivation, and place through the authored experiences of storytellers.

Theoretically speaking, narrative is immediately useful to a variety of tasks involving imagery. For one thing, narrative is important to the process of *ekphrasis,* or the "verbal representation of visual representation" (Mitchell 152). We are interested in this other direction, however: the communication of narrative through, not by, <img> and its attributes. This narrative praxis is sometimes realized through what Barthes describes as the relay function of linguistic messages ("Rhetoric" 41). Here, image and text complement one another, telling a story through the synergistic relation of the visual and verbal modes. Barthes notes that the typographic and visual modes are not commensurate: "It is the image which detains the informational charge and, the image being analogical, the information is then 'lazier.' . . . The costly message and the discursive message are made to coincide so that the hurried reader may be spared the boredom of verbal 'descriptions,' which are entrusted to the image, that is to say a less 'laborious' system" ("Rhetoric" 41). Barthes's comment on efficiency is important because it suggests that imagery has a more immediate payoff and faster transportability than verbal descriptions.

In addition to narrative's ability to translate elements of the visual into verbal texts, we can also consider the ontological dimensions of this form. Psychologically speaking, narrative is believed to be an integral part of human development, memory, and cognition. Even from early childhood, humans use the narrative form to encapsulate their experiences and to arrange these experiences in appropriate cognitive structures (often referred to as a scripts or schemas) with means for representing temporal and causal relationships (see Schank; Bruner; Herman). H. Porter Abbott explains the linkage between narrative and childhood development:

> Narrative capability shows up in infants some time in their third or fourth year, when they start putting verbs together with nouns. Its appearance coincides, roughly, with the first memories that are retained by adults of their infancy, a conjunction that has led some to propose that memory itself is dependent on the capacity for narrative. In other words, we do not have any mental record of who we are until narrative is present as a kind of armature, giving shape to the record. (2–3)

On a grander historical scale, narrative imagery served as one of our earliest forms of communicative technology, with visual stories and story fragments operating as devices to convey information and observations about the world and its behaviors for preliterate societies. The Chauvet cave paintings, discovered in 2004, depicted over 400 animals including lions, rhinos, bears, and panthers, and, remarkably, were dated to be over 30,000 years old (Faigley et al. 2). Although there is no clearly discernible story found in these paintings, there is certainly an assemblage of raw materials (environments and narrative agents) from which to assemble a multitude of plots and stories of various types (e.g., cautionary tales, celebratory epics, illustrative texts about how and when to hunt certain animals).

The ability of imagery, especially single images, to tell stories seems unusual at first. Mitchell further explains this as a misleading effect of believing that a particular mode of language must also approximate the condition in which it is inscribed in a medium. He writes:

> We think, for instance, that the visual arts are inherently spatial, static, corporeal, and shapely; that they bring these things as a gift to language. We suppose, on the other side, that arguments, addresses, ideas, and narratives

are in some sense *proper* to verbal communication, that language must bring these things as a gift to visual communication. But neither of these "gifts" is really the exclusive property of their donors: paintings can tell stories, make arguments, and signify abstract ideas; words can describe or embody static, spatial states of affairs. (160)

In terms of narrative and the online experience, the same capabilities that narrative provides in the physical world—namely, the establishment of an intersubjective cognitive space in which people can relate to the images, experiences, and descriptions provided by others' stories—are also critical in virtual space. Further, we speculate that for online communication, the linkage of imagery and narrative may work together in order to better facilitate a rich shared understanding of places in cyberspace. If we are to accept the prior definition of narrative as an expression of environment, character, and plot, then we must begin to recognize the ways in which these various elements are produced and arranged through online imagery. Furthermore, we must consider the psychological constructs of images in space as they contribute to or mediate the shared values of a community.

Conceptualizing Image and Narrative for the Web

Ruth Wajnryb proposes that one might think of experience as the raw material of story, and story as the raw material of narrative text (8). In this regard, one might form many stories from a core set of experiences, each of which may be expressed through a different medium. For instance, a conversational face-to-face story would be expressed as a narrative text through the medium of air, while a work of fiction might refashion or reformulate experiences for creative effect within the boundaries of printed pages (and, ultimately, of the bound book). Although the same sets of experiences are being used in each instance, the particular representations and effects of these experiences depend on the context of the communicative act as well as on the intent of the author. Thus, Wajnryb's model can be represented as *experience → story → narrative text.*

In order to extend Wajnryb's model so as to position narrative texts as shared intersubjective spaces, it is useful to once again return to Barthes. Barthes's concept of the image repertoire situates shared identity as a merging of personal and social space as interpreted by an individual. He describes the concept of image-repertoire as follows: "It is the discourse of others *insofar as I see it* (I put it between quotation marks). Then I turn the scopia on myself: I see my language *insofar as it is seen:* I see it *naked* (without quotation marks): this is the disgraced, pained phase of the image-repertoire" (Barthes, qtd. in Kopelson 59). The image repertoire is therefore both intuitively natural, in our observation of others, and awkwardly introspective, in our analysis of our own discourse through the eyes of others. Through the use of images, though, this awkwardness is reduced; no longer are there so many potential signifiers—as noted previously, the image is ideally suited for lazy readers. In this case, the awkwardness only emerges before the construction of an image, as in the selection of a suitable picture for one's identity (will they understand why I chose Frodo for my avatar?) or a photographic pose. Barthes considers this plight: "No doubt it is metaphorically that I derive my existence from the photographer. But though this dependence is an imaginary one (and from the purest image-repertoire), I experience it with the anguish of an uncertain filiation: an image—my image—will be generated: will I be born from an antipathetic individual or from a 'good sort'?" (*Camera Lucida* 11). As every individual must wrestle with this forced duality, online images can become a safe way to create spaces in which a larger granularity of identity is encouraged by cultural practices within that space. One might say, "It is not necessary to define myself wholly, but only in parts salient to this Web community. At this site, all individuals are automobile aficionados, so which vehicle avatar will I choose?" Or, "Within this Web ring, individuals share photomontages of their families, so which images will I choose to communicate the essence of my family's values?" The construction of a shared image-repertoire can serve as an important first step in priming an online community for a shared sense of narrative place.

We contend that the `<img>` tag facilitates narrative comprehension through several mechanisms, including an efficiency of interpretation in terms of online content, the creation of a shared image-repertoire, and the capacity for connoted nuance. Imagery online democratizes the process of communication, enabling a diverse array of individuals to communicate quickly and easily and to develop a shared understanding of hypertextual messages. Our suggested model is represented as the reformulation of Wajnryb's model as follows: *experience* → *story* → *images and texts* → *shared image repertoire* → *shared narrative.*

A shared narrative is a vehicle for communal telepresence. Consider an author (or Web developer, in this case) who chooses representative images that she feels are rhetorically appropriate for her Web site. These images may include representational imagery, or they may be more abstract. They may be numerous or few in number. Even the images that make up the structure of a Web site contribute to viewers' impressions, attitudes, and thoughts about that site. The goal of the developer is to create a site in which her audiences are predisposed to think of that Web space in a common way. This may be a rhetorical goal, as in persuading readers to buy a product or spend money for services, or a psychological goal, as in asking visitors to feel comfortable sharing intimate moments of their lives with like-minded individuals. Imagery can assist with these goals in a variety of ways. For a rhetorical goal, one might use soothing colors, neutral graphics, and a professional or corporate aesthetic that brings to mind prior experiences with successful online transactions. For the psychological goal, something as simple as providing snapshot photographs of community members can do a great deal to convince people to open up and share their own stories.

The critical element to note here is the modulation of perception and cognition as facilitated by the image content. Here, individual experiences are directed to specific focal points (particular images as displayed using the <img> tag on the Web) in turn, focusing these experiences on specific aspects of one's own background (and any associated cognitive schema engaged through that introspective process). These focal images can then normalize or focus attention on certain cues presented within the environment, thereby enhancing the sense of a shared story within a community. Perceptions are reinforced through the mechanisms of the narrative text, which in this case is the Web. Furthermore, two critical elements of narrative comprehension can be easily stimulated by the use of imagery: causality and temporality. Both can be observed in the positioning of the present with past experiences in either virtual or physical spaces: a reader sees a friendly face displayed on a customer's review profile, the reader associates friendly faces with prior positive transactions, and this cognition causes the reader to make a new purchase on the basis of the friendly ethos engendered by the site's use of embedded images. This is congruent with research from the field; in her analysis of three different credibility studies, Barbara Warnick (262) writes that despite popular belief, users often evaluate the credibility of Web sites based on factors such as professionalism of design and usability more than on authorial credentials.

It is certain that written texts can have as rich and well-developed narratives as images, video, or other forms of media. Our point is that the characteristics of visual media provide the following benefits of creating and sustaining a shared narrative of a particular type (e.g., personal narrative, ideological narrative, organizational narrative):

SPEED: This is the "a picture is worth a thousand words" phenomenon. Simply put, visual media are able to communicate complex narrative information more quickly than textual media, particularly because additional words, sentences, paragraphs, or even pages must sometimes be read in order to wholly engage with the causal structure of a plot or the descriptive elements of a character. Less supporting information is necessary to support a visual story. For example, the paragraph describing the physical features of a protagonist could be replaced with an image revealing these features either explicitly through dimension and coloring or implicitly through posture, gesture, or subtle facial expressions. The trade-off here is often one of imagination; while texts allow for more of an imaginative bridge to form between a reader and a work, the bridge between a visual piece and its viewer is arguably somewhat narrower.

FOCUS: What a visual narrative loses in imaginative potential, it gains in focus. A certain degree of focus is desirable and necessary for shared narratives, and the shared point of reference enabled through forms of image are critical for narrative cohesion. In other words, the inclusion of imagery often helps ensure that the many diverse (and often impatient) Web users key in on similar aspects of the narrative.

SUBTLETY OF DETAIL: Although textual narratives are quite good at communicating subtlety and connoting the layers hidden beneath the text, imagery provides its own mechanism of connotation and conveyance of affect through the use of color, artistic technique, and style. For instance, a Gothic narrative might effectively be conveyed by using a dark color palette and a degree of solemnity rather than through explicit items or objects included or excluded from the canvas. This is particularly useful on the Web, where a site may need to communicate underlying themes implicitly. For example, consider an online banking Web site; they cannot dedicate textual space on every page to assure visitors that they are safe and trustworthy (nor is it likely that modern Web users would take the time to read such a statement if it were included). Instead, effective online banking sites use imagery to signal these nuances about their company's narrative.

Daniel Stokols and Maria Montero relate these properties specifically to the Internet:

> Whereas non-Internet forms of communication (e.g., reading a book, watching TV, talking with others on the telephone, or corresponding with them by

surface or air mail) can bring geographically distant people and places psychologically closer to the individual, the Internet differs from these other
media in some important respects. . . . Internet-based communications
often combine textual, graphic, and auditory modalities (e.g., real-time video
images of the people one is communicating with as well as dynamic views
of their physical surrounds). Printed media are quite capable of depicting
faraway people and places through photographs, drawings, and text, but
they do not provide real-time interactive views of distant people and events;
nor can they deliver nearly instantaneous, multimodal communications.
(663–64)

Following this logic, we argue that images have more immediate information than texts simply because they are able to more rapidly capture
and transmit holistic detail. Furthermore, images (and other multimedia)
have unique power in the information-complex modern environment.
They facilitate the quick conveyance of detailed information, and they help
people from diverse backgrounds more easily develop a shared focus and
shared narratives.

Next, we should consider the psychosocial effects of imagery on the
Web. As stated previously, we believe that imagery both democratizes and
focuses the process of online storytelling, enabling more people to tell
stories through the medium while at the same time focusing and modu-
lating perceptual cues in order to formulate a shared narrative and a shared
rhetorical experience. In cyberspace, common narratives that are attached
to Web sites (and their content) allow individuals to develop an intersubjective rhetorical area encircling those virtual spaces. These resemble the
shared narratives people build around particular locations in the real world.
Thus, we argue that once the conceptualization of virtual space is shared
among the community of users, the space takes on many of the characteristics that would be found in physical locations.

From Cyberspace to Cyberplace

The Web can be thought of as a medium where information is conveyed
in narrative form. It can also be considered a narrative object in its own
capacity or rather as a series of narrative objects—a network of hypermedia sites that have the capacity to take on causal and temporal meaning
and become integrated into viewers' schema. However, this description is
a bit cumbersome. Thus, following the work of Mark Nunes and Barry

Wellman, we suggest that the Web be considered as a collection of places or be conceptualized as a collection of cyberplaces rather than as a singular cyberspace. In other words, we suggest that shared online narrative can be thought of as a shared sense of place, and likewise the notion of place is itself a particular idiosyncratic narrative form. If we recall memories of places that are special to us in the real world, they often arrive to us in narrative form. We visited these locations at a particular time and within a particular geographical environment, we ourselves or someone close to us played an active role in making the place important, and there was some central concern or "plot" that led us to visit that location in the first place. There may be obstacles or barriers preventing us from visiting that location as often as we might like, even if these barriers are as mundane as lack of time, fiscal resources, or an emotional adversity of some sort. Similarly, the memory is situated in both temporal (at the very least in regards to things that happened before and after our encounter with the place) and causal (the cause and effect relationships that led us to that place) relationships. With physical places, then, we often have all of the prototypical elements of story: time, place, conflict/drama, plot, causality, and temporality.

The benefit of conceptualizing the virtual Web as a series of places is twofold. First, the notion of place is intuitive; it is a metaphor to which most everyone can relate. Second, the concept of place has been extensively researched in a variety of academic disciplines since the 1970s. As a result of those investigations, useful approaches, constructs, and ways of thinking about place have been offered, debated, and reviewed. These dialogs can now be leveraged in hypertext studies; although some traditional concepts may be changed by the translation into new domains, they can still serve as useful conventions and talking points for a new generation of dialog. One area that offers some useful ideas is that of environmental psychology.

The study of place as a psychological construct is extensive. Edward Relph and David V. Canter have each authored comprehensive books entitled, respectively, *Place and Placelessness* and *The Psychology of Place.* More recently, Irwin Altman and Setha M. Low edited *Place Attachment,* which offers a rich discussion on the topic. Michael E. Patterson and Daniel R. Williams's article includes a summary of the academic debates of place and a review of recent research, and Lynne C. Manzo's review article discusses the connection between place and emotionality.

In this line of research, place is traditionally defined as a physical space that is imbued with meaning (Low and Altman 5). As Yi-Fu Tuan explains, "What begins as undifferentiated space becomes place as we get to know it better and endow it with value" (6). Because of this, place is inherently subjective. A place develops because people assign meanings, define appropriate behaviors, and collectively construct the essence of it. Consider the notion of home. A home may consist of a physical structure, but for many people, *home* is defined by a narrative, consisting of memories, emotions, and ideas, sentimental experiences, comfort, and social identity. For some, home may not even comprise a geographic location, but instead it may be purely narratological (Cheng, Kruger, and Daniels 89), a fleeting idea that can only be conceptualized with the participation of other agents (e.g., family members or friends) or activities (e.g., eating favorite meals). In this example, home is more than a physical location or an assemblage of physical features. The essence of that place, or the larger notion of sense of place, is an emergent property derived from the interplay of perceptual cues, behavioral norms, and social meanings, and "hence, place is not simply an inert container for biophysical attributes; place is constructed—and continuously reconstructed—through social and political processes that assign meaning" (Cheng, Kruger, and Daniels 90).

Two competing paradigms are commonly used to understand place: phenomenology and positivism (compare Patterson and Williams; Cheng, Kruger, and Daniels; Stedman). Phenomenology stems from a philosophical stance that reality is inherently subjective. As Patterson and Williams, quoting Amedeo Giorgi, explain, "this paradigm is concerned with presences (or objects) as they appear in consciousness. That is, objects are not of interest in terms of their 'objective,' 'real,' or 'existential' sense; rather the focus is on the meaning 'of the object precisely as it is given' to an individual" (369). In contrast, the positivist (also called psychometric) tradition considers reality tangible, understandable, and measurable. Positivistic research attempts to be definitive and consequently relies on quantitative methodologies, structured definitions, and scientific hypothesis testing (Patterson and Williams; Stedman; Hidalgo and Hernández).

In regard to sense-of-place research, the phenomenological perspective sees places as socially constructed symbols that hold power over people. This approach eschews efforts to define the specific mechanisms that create the meaning of a place. Instead efforts focus on understanding the place's influence on people's behaviors, attitudes, and ideas (Cheng, Kruger, and Daniels 91–92). Relph explains that the sense of place "is not just a formal concept awaiting precise definition. . . . Clarification cannot be achieved by imposing precise but arbitrary definitions" (4). From the positivistic perspective, the sense of a place is derived from individuals' cognitive strategies (e.g., mental schema, heuristics), which are influenced by their personalities and social/cultural norms. The cognitive approach is mainly interested in understanding how people categorize places and how individuals develop the cognitive strategies used for classification (Cheng, Kruger, and Daniels 92).

For introductory purposes, this description should serve. However, Patterson and Williams offer a much more detailed survey of the field. Their model examines the conceptual origins, underlying assumptions, and research programs associated with several disparate veins of place research.

In summary, place has been studied from a variety of psychological perspectives. Places are generally considered centers of meaning, which are constructed of physical features and of social and psychological processes. Many years of academic inquiry have shaped our current understanding of place, and we believe that these efforts can now be applied to the study of hypermedia places.

Virtual Places

Although some have argued that place becomes increasingly irrelevant in the modern information-technology-mediated world (Meyrowitz; Giddens), many researchers have embraced the idea of virtual places (Stokols and Montero; Blanchard; Wellman). Proponents of virtual places conclude that individuals who use computer-mediated communication can actually feel a sense of place, which develops from the combination of social attachments and conceptual models of virtual space. For instance, Howard Rheingold argues that after a virtual community has reached a threshold number of shared experiences, "the community takes on a definite and profound sense of place in people's minds" (64). Anita Blanchard echoes this perspective, and she suggests that traditional approaches to understanding (physical) places be applied to virtual places. Furthermore, Stokols and Montero outline an entire research agenda for applying environmental psychology to the Internet. As they explain, "The field of environmental and ecological psychology thus provides a useful background for developing a conceptual analysis and programmatic agenda for future research on the ways in which the Internet and Web are transforming the quality and structure of people-environment transactions" (666).

We can use this additional theory from environmental psychology to augment our proposed model, described previously. The sequence thus takes on a new dimension: *experience → story images and texts → shared image repertoire → (a particular narrative) sense of place.* This is a useful conceptualization because now many traditional people–environment concepts may effectively translate from tangible to virtual places. We next consider a few environmental research concepts and examples that are particularly relevant to our discussion. This discussion demonstrates the power of the cyberplace metaphor and its applicability to hypermedia research using particular concepts derived from research in environmental psychology and analyzed from the framework of rhetorical studies.

Imageability, Legibility, Wayfinding, and Image Schemata

Environmental psychology offers some useful applied techniques for thinking about the construction of narrative `<img>`-texts. First, consider imageability, a construct that describes how individuals experience places. Kevin Lynch originally defined imageability as the quality of a physical place that gives it high probability of evoking a strong image in any given observer (2). He believed that imageability could be defined by historic, personal, and social meanings, as well as five physical elements: paths, nodes, edges, districts, and landmarks. Imageability can be thought of as a zeitgeist for a physical location, or in other words, the collective spirit of a place. In virtual space, we can appropriate the use of this term to describe the ways in which virtual cues trigger these historic, personal, and social meanings. Further, we can use the conceptual elements of paths, nodes, and edges to define the structural characteristics of hypermedia, the hyperlinks and hypertext files that make up the Web.

To define a collective place, then, is to properly situate a site within the appropriate context of districts, which may be linked Web sites. The use of appropriate imagistic landmarks further focuses perception and allows multiple visitors to form common mental images and to use common vocabularies when defining virtual places. These images all contribute to the formation of shared narratives that often mirror experiences in the real world. For example, if people are familiar with Amazon.com, then they can use another e-commerce site more easily because its visual design is likely to resemble Amazon.com, and thus the visual cues are likely to rapidly activate previously stored scripts and schemata for shopping. The emergent narrative in this example, although not always entertaining in the sense of a novel or a film, is nonetheless critical for cognitively engaging the character (here, Web users) with a sense of purpose, offering cues for appropriate behaviors, and evoking cognitive and emotional reactions. All of this is encapsulated by the sense-of-place perspective. Whether for entertainment, commerce, communication, or simply business, the visual strategies of contemporary Web sites often implicitly or explicitly function in a rhetorical sense to encourage this sense of narrative activation and to share and sustain this narrative with other visitors. In other words, the images online—from individual photographs to blog background designs instituted by a community or organization—each contribute to the development of a distributed, intersubjective narrative, which can be conceptualized as place.

As a second example, consider the rhetorical function of an informational site designed to connect various audiences within an organization. For example, the Web site shown in Figure 12.1 was designed by one of us (S.L.S.) to provide university students, faculty, staff, and administrators with curricular and scheduling information for an academic unit. The imageability of this site suggests an amalgamation of new and old, modern

Figure 12.1. Web site designed by author S.L.S. for University of Central Florida Digital Media Program.

and traditional. Three primary districts are defined: an upper boundary that defines an institutional ethos of "emerging media" (complete with a gentleman sporting pink hair in the color version of this screenshot), and two lower districts with images of a more traditional technology (calendar) in the lower left quadrant and a more recent technology (computer keyboard) in the lower right. Collectively, these images combine to suggest a spirit of innovative yet organized activities. One can then consider the shared image repertoire made possible by the imagery on this Web site; this leads to potential narrative fragments of plot (what might it be like to be a student in the program?) or character (what are others' perceptions of the program? What it might be like to work as a faculty member teaching classes for that program?). In this case, the persuasive goals of the site are supported (or perhaps undermined, if one does not identify well with pink hair) by the imageability provided by these districts and nodes provided by the <img> tags. This example is also one in which a sense of place is not cocreated by an emergent community, but rather carefully considered in order to serve an existing community, as many Web sites are designed to do.

Along with imageability, Lynch describes the idea of legibility: "the ease with which its [the cityscape's] parts can be recognized and organized into a coherent pattern" (2). Lynch is tauntingly close to applying environmental studies to media studies: "Just as this printed page, if it is legible, can be visually grasped as a related pattern of recognizable symbols, so a legible city would be one whose districts or landmarks or pathways are easily identifiable and are easily grouped into an overall pattern" (3). Successful place building, or at least successful narrative schemata activation, is largely informed by a developer's ability to choose legible images, or images that evoke desirable feelings and emotions across audiences that may be demographically dissimilar. But also, and more importantly, imageability relates to the pattern of images, their summative qualities and contrasting forms. It is this gestalt, or patterned image, that informs the identity of a place, or for our purposes, the shared narrative of a cyberplace (Lynch 83–90).

Part of legibility also involves being able to project oneself into the virtual medium in a way such that landmarks and online imagery become meaningful for a particular task. This is the introspective function of the image repertoire. Another example helps to illustrate this. Consider the common discursive act of writing a collective journal for one's family or friends. For example, if one were to design a family journaling site, it would make sense to construct a virtual place in which various family members served as focal points through which to voyage chronologically through different stories and events. In Figure 12.2, a personal Web site for one of the authors (R.M.) uses imagery to project a sense of moving through a young child's life in journal form. Although the entries themselves are primarily textual, the omnipresent banner provides focus and consistency to this experience while subtly reinforcing the theme of the site, which is a journey of discovery for a young child as seen from the perspective of his parents. Here speed, focus, and subtlety of detail are all provided through a simple <img> banner. Similarly, other visitors within this particular discourse community (parents of young children) can quickly focus on the types of stories that are likely to emerge from this Web place and decide whether or not they would like to contribute to the community. In this case, a sense of place is both designed for others and potentially shifted into new organic forms on the basis of community participation from individuals identifying with this image.

Next, we can consider the idea of wayfinding. Lynch explains, "In the process of way-finding, the strategic link is the environmental image, the generalized mental picture of the exterior physical world that is held by an individual. . . . The need to recognize and pattern our surrounding is so crucial, and has such long roots in the past, that this image has wide practical and emotional importance to the individual" (4). Jerry Weisman also notes the particular importance of architectural differentiation for wayfinding. In most analyses of wayfinding, sensory cues—particularly visual ones—are integral and personally meaningful facets of the process. Just as we use wayfinding to locate visual cues when driving to a new location in physical space, so do we use wayfinding to activate stored schemata and navigate through the complex information-rich nodes of the Web. In electronic space, the trick is to use rhetorical strategies that minimize the complexities of the real-world data—in other words, to suppress the

Figure 12.2. Author R.M.'s family journal.

overpowering feelings of information overload that would inevitably surface if all of the paths, nodes, and edges were to be revealed to a visitor of even a medium-scale Web site.

Returning to our narrative analogy, one might consider wayfinding as a technique for accessing given paths through a complex narrative environment, or for jumping ahead to critical plot points of stories already in progress. Wayfinding can be used for pattern matching in image repertoires when common themes are associated with different images. Again, cultural conventions and prior experiences assist with this. For example, the virtual equivalent of the checkout line in a supermarket is the shopping cart icon; in these instances, one knows that clicking on this icon will propel one into the culminating event of a particular story—the purchase of a material item from a virtual storefront. Although the specific look and feel of these checkout mechanisms will differ across sites, the narrative sense of place is still the same: a virtual environment is created in which to complete the transaction and to feel secure providing credit card information and other personal details to a machine rather than to another human being.

Narrative devices can also provide the context for a story, which relates to the notion of the patterned image. The patterned image suggests certain organizational narratives, which in turn convey ideological or institutional themes. In some cases, connective stories can be told from the simple choosing of an artistic style or genre that evokes nostalgia or pleasant feelings about one's experiences with another medium. For example, Figure 12.3, which depicts a banner image designed by Zach Whalen for a themed conference, World Building: Space and Community, depicts a game-stylized aesthetic that was used to communicate the conference's theme of space and community in online environments. The image is present in the Web medium, but it calls forth memories of video gaming media. In this case, wayfinding functions as a conceptual rather than a spatial bridge. In other words, the image connects visitors not with other locations

Figure 12.3. Web site designed for the World Building: Space and Community conference.

on the Web, but with prior experiences playing video games such as *The Sims* or *Sim City*, in which this type of visual aesthetic might be encountered. In doing so, the primary conference theme is reinforced and the institutional narrative of fruitfully combining virtual and physical space for scholarly considerations is subtly passed along to potential participants and visitors of the site.

A final concept useful to place making, and one related to wayfinding, is that of image schemata. Image schemata are metaphors people use to comprehend experiences while moving through and interacting with an environment (Johnson 28–30). In other words, they are the "recurring, imaginative patterns" (Frank and Raubal 69) through which people interpret the physical world. "An image schema can, therefore, be seen as a very generic, maybe universal, and abstract structure that helps people to establish a connection between different experiences that have this same recurring structure in common" (Raubal et al. 5). Mark Johnson, a pioneer of image schemata theory, suggests several prototype schemata. For instance, CONTAINERS, consist of an inside, an outside, and a boundary; things can enter a container (go *in* a building), leave a container (get *out* of a car), or cross a container's boundary (pass *through* a doorway).

Johnson's work helped formalize these commonsense patterns that people implicitly use to understand space, and several scholars have since suggested that image schemata might be used for interface design. For instance, "Image schemata are considered good candidates as a foundation for the formal definition of spatial relations. [Werner] Kuhn [a professor of geoinformatics and a prolific author] has pointed out the importance of image schemata as a tool to build 'natural' (i.e., cognitively sound) user interfaces" (Frank and Raubel 3). We suggest that image schemata may also be a useful tool for thinking about the relations that exist in virtual place-making narratives.

For example, a popular activity today is to participate with social networking sites such as Flickr, a photo-sharing Web site that is particularly relevant to our discussion of images. Flickr allows individuals to post images and their textual descriptions, review others' photos, form communities of friends and acquaintances, and interact with those discourse communities. Consider this scenario: A person logs *in* to her Flickr site, then clicks *on* a friend's personal photo, which *links to* her friend's page. The use of *in* evokes the CONTAINER schema and implies that the virtual place possesses a conceptual inside, outside, and boundary. "On" suggests a special type of object, called a GATEWAY schema. "Links" reflects the LINK schema that, not surprisingly, connects objects in time or space (here connecting one page to another). Finally, "to" suggests the use of the PATH schema, an especially important wayfinding device that is defined by a starting point, end point, and connection between the two.

This is a simple application of image schemata theory to cyberspace, and even without images, the Internet would evoke such metaphors. However, with the benefit of the <img> and the composite designs it builds, more complex abstractions emerge. For instance, once our Flickr user reaches her friend's page, she can click *on* a collaged icon that represents a *set of photos*. Once the link is pressed, our user *moves deeper in* to the photo set where she can see a *collection* of associated image *objects*. She can then click *on* a single photo and drill *down to* its stand-alone page. This small description already implies many sophisticated schemata, including COLLECTION, CENTER–PERIPHERY, NEAR–FAR, and SPLITTING. Overall, the site's basic visual design (i.e., its image pattern) suggests a sense of space, evokes specific spatial metaphors, and facilitates the formation of a strong cognitive representation of Flickr as a tangible space.

The personal narratives interwoven with the user-posted photos construct the meaning, social context, and emergent sense of place within Flickr, or the image-repertoire of the community. Thus, at a macro level, Flickr is a space that can be described by the metaphors of image schemata, similar to a large city. At a micro level, Flickr comprises places constructed by individuals and communities who post, view, discuss, and otherwise interact with each others' personal narratives. Like a well-known neighborhood near one's home, this localized Flickr is defined both by the metaphors of image schemata as well as by personal narratives imbued within that space. As such, a shared sense of space and then of place is fostered by the <img>-texts.

Conclusion

As this discussion has demonstrated, many psychological concepts associated with the study of place can be beneficially applied to media studies. We hope that hypermedia scholars will see the similarities and synergies between traditional environmental psychology research and today's studies of hypermedia. We believe that the traditional study of place may be effectively translated, and that the conceptualization of cyberplace is a strong foundation on which to construct and communicate relational theories of hypermedia. Furthermore, the concepts of imageability, legibility, wayfinding, and image schemata are useful tools for the construction of shared <img>-text Web sites with the potential for rich narrative histories and stories supported by various communities of individuals. Images make it easier for individuals to ascribe meaning to the Web, not just to fragmented nodes of text and dissociated data, but to virtual locations, Web sites, and the individuals and communities of people who frequent them. The narratives that individuals construct about the Web can be conceptualized through the study of place, which understands physical locations on the basis of the narratives (and the psychological outcomes caused by the narratives) that individuals create for physical spaces. Images online are equivalent to physical/spatial cues in real-life settings. By conceptualizing the Internet as a place and virtual locations "on" the Internet as places, it becomes easier to understand the cues important for the narrative construction of virtual experience. This viewpoint suggests exciting new areas of research for hypermedia studies.

Works Cited

Abbott, H. Porter. *The Cambridge Introduction to Narrative.* Cambridge: Cambridge University Press, 2002.

Altman, Irwin, and Setha M. Low, eds. *Place Attachment,* Human Behavior and Environment: Advances in Theory and Research. New York: Plenum Press, 1992.

Barthes, Roland. *Camera Lucida: Reflections on Photography.* Translated by Richard Howard. New York: Hill and Wang, 1981.

———. "Introduction to the Structural Analysis of Narratives." In *Narratology,* edited by Susan Onega and José Ángel García Landa, 45–60. London: Longman, 1996.

———. "The Photographic Method." Translated by Stephen Heath. In *Image-Music-Text,* 15–31. New York: Hill and Wang, 1977.

———. "Rhetoric of the Image." Translated by Stephen Heath. In *Image-Music-Text,* 32–51. New York: Hill and Wang, 1977.

Blanchard, Anita. "Virtual Behavior Settings: An Application of Behavior Setting Theories to Virtual Communities." *Journal of Computer-Mediated Communication* 9 (2004). Accessed 2 August 2010. http://jcmc.indiana.edu/vol9/issue2/blanchard .html.

Bruner, Jerome. "The Narrative Construction of Reality." *Critical Inquiry* 18 (1991): 1–21.

Canter, David V. *The Psychology of Place.* London: Architectural Press, 1977.

Cheng, Antony S., Linda E. Kruger, and Steven E. Daniels. "'Place' as an Integrating Concept in Natural Resource Politics: Propositions for a Social Science Research Agenda." *Society and Natural Resources* 16 (2003): 87–104.

Faigley, Lester, et al. *Picturing Texts.* New York: Norton, 2004.

Flickr. Accessed 31 December 2008. http://www.flickr.com/.

Frank, Andrew U., and Martin Raubal. "Formal Specifications of Image Schemata for

Interoperability in Geographic Information Systems." *Spatial Cognition and Computation* 1 (1999): 67–101.

Garrett, Jesse James. "Ajax: A New Approach to Web Applications." *Adaptive Path.* 2005. Accessed 8 October 2006. http://adaptivepath.com/.

Giddens, Anthony. *Modernity and Self-Identity: Self and Society in the Late Modern Age.* Stanford, Calif.: Stanford University Press, 1991.

Giorgi, Amedeo. "The Theory, Practice, and Evaluation of the Phenomenological Method as a Qualitative Research Procedure." *Journal of Phenomenological Psychology* 28 (1997): 235–60.

Google. *Google Maps.* Accessed 11 April 2007. http://maps.google.com/.

Herman, David. *Narratology as a Cognitive Science.* 2000. Accessed 19 December 2006. http://www.imageandnarrative.be/.

Hidalgo, M. Carmen, and Bernardo Hernández. "Place Attachment: Conceptual and Empirical Questions." *Journal of Environmental Psychology* 21 (2001): 273–81.

Johnson, Mark. *The Body in the Mind: The Bodily Basis of Meaning, Imagination, and Reason.* Chicago: University of Chicago Press, 1987.

Kopelson, Kevin. *Neatness Counts: Essays on the Writer's Desk.* Minneapolis: University of Minnesota Press, 2006.

Low, Setha M., and Irwin Altman. "Place Attachment: A Conceptual Inquiry." In *Place Attachment,* edited by Irwin Altman and Setha M. Low, 1–12. New York: Plenum Press, 1992.

Lynch, Kevin. *The Image of the City.* Cambridge, Mass.: MIT Press, 1960.

Manovich, Lev. *The Language of New Media.* Cambridge, Mass.: MIT Press, 2001.

Manzo, Lynne C. "Beyond House and Haven: Toward a Revisioning of Emotional Relationships with Places." *Journal of Environmental Psychology* 23 (2003): 47–61.

Meyrowitz, Joshua. *No Sense of Place: The Impact of Electronic Media on Social Behavior.* New York: Oxford University Press, 1985.

Mitchell, W. J. T. *Picture Theory.* Chicago: University of Chicago Press, 1994.

Nunes, Mark. *Cyberspaces of Everyday Life.* Minneapolis: University of Minnesota Press, 2006.

Patterson, Michael E., and Daniel R. Williams. "Maintaining Research Traditions on Place: Diversity of Thought and Scientific Progress." *Journal of Environmental Psychology* 25 (2005): 361–80.

Raubal, Martin, Max J. Egenhofer, Dieter Pfoser, and Nectaria Tryfona. "Structuring Space with Image Schemata: Wayfinding in Airports as a Case Study." In *Proceedings of the International Conference on Spatial Information Theory, Laurel Highlands, 15–18 October 1997,* edited by Stephen C. Hirtle and Andrew U. Frank, 85–102. London: Springer-Verlag, 1997.

Relph, Edward. *Place and Placelessness.* London: Pion, 1976.

Rheingold, Howard. *The Virtual Community: Homesteading on the Electronic Frontier.* Reading, Mass.: Addison-Wesley, 1993.

Schank, Roger C. *Tell Me a Story: Narrative and Intelligence.* Evanston, Ill.: Northwestern University Press, 1995.

Stedman, Richard C. "Toward a Social Psychology of Place: Predicting Behavior from

Place-Based Cognitions, Attitude, and Identity." *Environment and Behavior* 34 (2002): 561–81.

Stokols, Daniel, and Maria Montero. "Toward an Environmental Psychology of the Internet." In *Handbook of Environmental Psychology*, edited by Robert B. Bechtel and Arza Churchman, 661–75. New York: Wiley, 2002.

Tuan, Yi-Fu. *Space and Place: The Perspective of Experience*. Minneapolis: University. of Minnesota Press., 1977.

Wajnryb, Ruth. *Stories: Narrative Activities in the Language Classroom*. Cambridge: Cambridge University Press, 2003.

Warnick, Barbara. "Online Ethos: Source Credibility in an 'Authorless' Environment." *American Behavioral Scientist* 48, no. 2 (2004): 256–65.

Weisman, Jerry. "Evaluating Architectural Legibility: Way-Finding in the Built Environment." *Environment and Behavior* 13 (1981): 189–204.

Wellman, Barry. "Physical Place and Cyberplace: The Rise of Personalized Networking." *International Journal of Urban and Regional Research* 25, no. 2 (2001): 227–52.

Whalen, Zach. "World Building: Space and Community." Presented at the Third Annual University of Florida Games and Digital Media Conference, 1–2 March 2007, Gainesville, Fla. Accessed 29 December 2008. http://worlds.gameology.org/.

WordPress. *WordPress Blog Tool and Web Platform*. 2007. Accessed 11 April 2007. http://wordpress.org/.

13   <table>ING THE GRID

Bradley Dilger

<table> *allows users to create tables within Web documents. With additional markup such as* <th>, <tr>, *and* <td>, *writers can specify headers, rows, and columns. Initially intended for representing data such as spreadsheets, tables were adapted for layout and design purposes. The first WYSIWYG Web authoring editor, Adobe PageMill, removed the laborious task of hand coding tables by allowing users to create them visually. While recognizing that tables were not intended to be design elements, a 2000 Macromedia tutorial for making tables in Dreamweaver 3 promoted* <table> *as a way to "give a designer control over the positioning of images and text on the page." The introduction of Cascading Style Sheets (CSS), however, made this usage obsolete. CSS organized space far more efficiently than* <table>. *In 2002, Web standards guru Jeffrey Zeldman proposed transitioning to CSS with a hybrid usage of tables and CSS for layout purposes. By 2009, tables were completely absent from his site, exemplifying the current movement away from tables as a tool for Web design.*

> Designers like systems. There is nothing like a system for creating a sense of control.
>
> **EMILY KING AND CHRISTIAN KUSTERS,** *RESTART: NEW SYSTEMS IN*
> *GRAPHIC DESIGN*

WHEN THE WEB EXPLODED IN POPULARITY in the early 1990s, the responsibility for building Web pages fell on a huge variety of people: programmers, prepress specialists, graphic designers, illustrators, even writers and editors. With few established best practices, design was fragmented. Overnight sensation David Siegel called for flashy "killer Web sites," and usability specialist Jakob Nielsen all but called for him to be killed. Many Web designers had no formal training, and the Web sites they created often mixed alignments, colors, typefaces, and graphical styles indiscriminately.

Furthermore, individual rendering of a given design varied much more widely than today, and small screen sizes, slow download speeds, and limited color palettes regularly obliterated designers' intentions. For professionals accustomed to using Aldus PageMaker, QuarkXPress, and Adobe Illustrator to produce printed texts that closely matched what appeared on their screens, this variation was unacceptable and unwelcome. For readers accustomed to the carefully regulated spaces of print, the Web was a mess—a mix of design philosophies and technological implementations, none of which were predictable.

Enter <table>, standardized as part of HTML+ by Dave Raggett and popularized by the Netscape Navigator browser. With <table> and a few more child tags, designers could divide pages into rectilinear containers with consistent sizes—that is, grids—replicating the appearance of the annual reports, brochures, and similar media they were accustomed to producing on screen and in print. Though <table> offered only limited control, it was a huge improvement, and takeup was swift; by February 1997, MIT, NCSA, NBC, and Yahoo had adopted the use of tables for layout on their sites. Designers everywhere began to use tables to create grid-like layouts, and competing browsers quickly adopted <table> and added enhancements. In late 1996, on the then-influential Web site Webmonkey, Derek M. Powazek wrote:

> I love tables. I know, I know. I'm a geek. But I really do.
>
> I remember when I first discovered tables, back in the days of Netscape 1.1. I stayed up all night and had a transcendental experience around 4 A.M. Finally, I could control where things fell on a page!

Powazek's joy in predictable layout of Web pages was part of a larger push-back, a move to gain control of the Web's chaotic design space. As Web-monkey contributor Jeffrey Veen wrote in a slightly different context, "As our networked world grows increasingly complex, layers and streams of information constantly bombard us. If you want to successfully design for the Web, you will take control of your content and boil it down to its very essence." Echoing this message was Nielsen, who in 1997 was emerging as the main authority on Web writing and design, tirelessly pushing the Web toward standardized design, content, and functionality. Layout grids built with <table> helped with all three. Their widespread adoption was part of a larger turn toward the international style, "a reductive, unadorned graphic design that aimed for objectivity and efficiency" (Middendorp 114). The style's popularity didn't hurt—the public knew Saul Bass's posters for films such as *Vertigo*, Allen Hurlburt's layout of *Look* magazine, and Paul Rand's corporate identities for IBM, Westinghouse, and ABC (all still in use today). For designers seeking to control the wild spaces of the Web, the international style was promising, given its central role of the grid, its preferences for high-contrast, sans-serif typography, and its close alliance with commerce and corporate identity.

Joe Clark suggests the name "international-compliant style" for this aesthetic, noting its spread is less a matter of "conscious decision" but rather being "committed to following official specifications." (The specifications Clark speaks of are those published by the World Wide Web Consortium [W3C] and supported by advocates of standards-compliant Web design, such as Jeffrey Zeldman.) In this essay, I extend Clark's argument by showing how the rise of this style was shaped by technical constraints as well as cultural factors. The widespread use of <table> as a layout device not only led to the development of a school of design, but it also influenced the development of "official specifications" themselves—particularly Cascading Style Sheets (CSS). Grids have helped transfer the apparatus of print to the spaces of the Web, serving as a strong counterpoint to the potentially destabilizing influences of associative logic, interactivity, and dynamic content. I conclude by discussing the close identification of grids with rationality and order, and the relationship of grids, minimalist Web design, and the emergence of protocological systems of control.

The International Style Comes Online

The story of the birth of the Web is well known: Tim Berners-Lee, a research scientist at the Organisation Européenne pour la Recherche Nucléaire (CERN), created a system for sharing publications with other scientists, and set up the first Web server and Web site on a NeXT workstation (Raggett et al. 21). Less well known: CERN is based in Switzerland, the home of Max Bill, Otl Aicher, Josef Müller-Brockman, and other designers for whom grids were central, if not essential. In their remarkable amount of design work for diverse American and European clients, codified in books such as Müller-Brockman's *Grid Systems in Graphic Design,* these designers developed an approach that focused on carefully designed typography set in orthogonal grids—indeed, for some, it is the international *typographic* style. As Jan Middendorp writes, the style attempts to "obtain maximum aesthetic effect by the simplest means: black and white photography, unjustified, homogeneous blocks of sanserif text, a conscious use of white space as a compositional element and the grid as a guiding principle" (114). Philip B. Meggs and Alston W. Purvis name the following stylistic components:

> A unity of design elements achieved by asymmetrical organization of the design elements on a mathematically constructed grid; objective photography and copy that present visual and verbal information in a clear and factual manner, free from the exaggerated claims of much propaganda and commercial advertising; and the use of sans-serif typography set in a flush-left and ragged-right margin configuration . . . mathematical grids [as] the most legible and harmonious means for structuring information. (356)

The remarkable fit between these visual elements and the technological capabilities of the early Web created strong correspondences between many early Web design styles and the international style. Above all, the early Web was about type: the Web began as a text-only space, and images weren't integrated with text until the 1993 release of the Mosaic browser. Though many Web sites adopted the flashing, multicolored, anything-goes approach identified with the Geocities Web service and parodied by Bruce Lawson, the technological limitations of HTML and the slow speed of analog modems supported a minimalist approach to Web design—a focus on text. Even sites that made extensive use of graphics frequently offered text-only versions. The international style supported this placement of type as the focus of layout, and sites that used <table> were able to approximate its gridded look very well.

As Meggs and Purvis describe it, the international style favored "sans-serif typography set in a flush-left and ragged-right margin configuration." So too on the early Web. Design pioneers adopted <table> to gain control over type, including Siegel, who used <table> for "margins and shorter line lengths" and "better typographic layout control" (6). However, even with CSS, "type on the web can be a mess" (Williams and Tollett), especially in comparison to the level of control designers can expect from page-layout software such as Adobe InDesign. And even today, Web browsers neither hyphenate nor justify type with good precision or flexibility (if at all). Whether using tables or CSS for layouts, many early Web designers chose left-justified type in order to deal with these limitations (Lynch and Horton)—a preference that endures.

Although Web pages are not built on a typographic grid, as is often the case with print designs in the international style, designers often used <table> to build designs that scaled relative to window and font size. In these "liquid" or "fluid" designs (Cederholm 3; Weakley), default type size is critical because pages are sized to fit their contents—a behavior that began with `<table>`. As users increase or decrease the text size, tables expand or contract accordingly. Wilson Miner's "Setting Type on the Web to a Baseline Grid" provides a method for strongly binding Web design and typography, so that "all the text on your page lines up across all the columns, creating a harmonious vertical rhythm." His well-received approach brings the focus on text from table to contemporary site-building methods.

Another tenet of the international style, "asymmetrical organization of the design elements," was characteristic of early Web pages, as complex alignment of multiple objects using only HTML is next to impossible without the use of `<table>` or other complicated markup (Niederst 168). Even using tables or CSS, it is easier to create an asymmetrical design than to balance elements predictably. Similarly, "elements on a mathematically constructed grid" have formed the core of many approaches to creating Web pages with sophisticated visual designs. Designers such as Siegel, who used `<table>` for more attractive typography, eagerly used it to divide pages into grids that emulate those used in graphic design. A standard of 600 pixels wide, matching the size of VGA monitors, emerged by early 1997. This fixed width led to predictability. As `<table>` was enhanced with the `<colgroup>` tag and other attributes, designers gained even more fine-grained control, with the ability to specify width of columns using absolute or relative measurements (pixels or percentages). Current methods for layout that rely on CSS often use absolute positioning, which literally places every page element on a mathematical grid, using X and Y coordinates.

These visual elements were by no means the only elements of the international style common to the minimalist look developed by early Web designers. Philosophies matched as well. Like the Swiss designers who advanced the international style, Berners-Lee, clicking away in their shadow in Geneva, strongly believed that good design and responsible use of technology would effect positive cultural change. They and he imagined texts and technologies working in harmony, with nearly endless possibilities in the operation of their modular grids. And like the Silicon Valley technologists who championed nascent dot-coms, arguing that they could restore the United States to its status as an economic superpower, the creators of the international style had imagined it as the path to postwar recovery for Europe. As Timothy Samara writes:

For the graphic designers who helped society struggle to move forward after two unimaginable wars, order and clarity became their most important goals. . . . Part of that order, of course, meant consumer comforts; and the businesses that provided them recognized soon enough that the grid could help organize their image, their corporate culture, and their bottom lines. (10)

Richard MacManus suggests that designers turned toward more refined visual styles to appeal to business as dot-com speculation waned in the late 1990s, and they sought to help their clients appear stable and reputable. But nobody did more to suggest that design and commerce went hand in hand than the aforementioned Jakob Nielsen. As Curt Cloninger writes, "his speedy download mantras and his least-common-denominator design approach" in which "the goal is effective communication—clear and uncluttered" (137) pushed hundreds of designers to look for an objective, efficient approach to design: exactly the values promised by the international style. Müller-Brockman had acknowledged the importance of beauty but insisted that "emotional" creativity must be tempered with intellectual and rational discourse, lest designs fail to achieve their necessary logical and systemic approach (160). Similarly, in his watershed *Designing Web Usability,* Nielsen was pragmatic: "While I acknowledge that there is a need for art, fun, and a general good time on the Web, I believe that the main goal of most web projects should be to make it easy for customers to perform useful tasks" (11). This approach was exemplified in "Concise, SCANNABLE, and Objective: How to Write for the Web," written with John Morkes. Offering the findings of a usability study, Morkes and Nielsen attack "marketese," suggesting writers strive for "straight facts" free of "promotional clutter." Without a doubt, they would support design that favored, as Meggs and Purvis suggest, "objective photography and copy" without the "exaggerated claims of propaganda." Given that Nielsen's influence is hard to underestimate—especially in the time period between 1997 and 2001—his repeated advocacy of the virtues championed by the international style provided a monumental boost to its popularity.

Nielsen also contributed to the broad turn toward the elements of print by suggesting adoption of the inverted pyramid used for news ("Inverted"). Other usability specialists suggested keeping content "above the fold," meaning important elements should fit in a single browser window (Spool). Navigational mastheads and footers, created with <table> cells using background colors, resembled the running heads and footers of print. Using <table> for layout provided a means for recreating commonly used print elements—columns, sidebars, pull quotes, and the like (Niederst 199). These elements were applied to whole sites by means of templating systems: nested table containers that contained areas into which content would be added—much like the pasteup boards used for newspaper and magazine production. Print journalism, with its column grid, applied consistently to every page, had become the model for the visual design of the Web. And standardization of visual appearance was matched by standardization of content. These visual identity standards or branding guidelines, including suggestions for writing on the Web derived from Nielsen and Morkes, were introduced by universities and other decentralized institutions that wished to unify their Web experiences.

A coherent style that embraced many of the tenets of the international style had emerged. Cloninger identified this clean look as "HTMinimaLism," naming it one of the most influential styles on the Web in 2001. He, too, recognized that its influence extended far beyond visuals, suggesting it was "the logical style for any site that has large chunks of text or a large catalog of displayable goods (read: e-commerce)" (160). Since Cloninger's writing, the presence of similar designs that favor minimalism and a grid-based approach has gradually increased. Prominent designers and programmers encourage their use (Vinh and Boulton; Atwood). Grids are prominent on many popular sites: Facebook, CNN, Wikipedia, and not surprisingly newspapers and news-oriented Weblogs such as the Huffington Post. Numerous Weblogs, content management systems, and Web application frameworks embed them in software. And the growing number of CSS tool kits that enable grid design—for example, Nathan Smith's 960 Grid System, Olav Bjørkøy's Blueprint framework, and the Yahoo YUI Grids CSS—all embody other characteristics of the international style as well. Today, the international style, with its grid-based layouts, minimalist approach, clean typography, efficient writing, and affinity for business, is the dominant graphic design style on the Web.

CSS: Built in the Shape of `<table>`

The technical limitations of HTML facilitated the Web's adaptation of a design style that matched the international style visually and ideologically. The minimalist, type-heavy, straightforward, business-friendly approach of the international style seemed natural because it was one of the few coherent design programs that could be effected with early HTML. As the Web matured, during the same time period I cover above, Håkon Wium

Lie, W3C style activity lead Bert Bos, and others working under the aegis of the W3C coordinated the writing of specifications that separated CSS from HTML and codified it as a language for visual design. Though the future of Web style was wide open at this point, the development of CSS was deeply influenced by `<table>`, and the designers and programmers who shaped CSS used tables for layout as a template. In this brief section, I document the ways the grid-based designs that epitomized the international style were transferred to technical specifications and best practices for CSS.

Although not directly associated with tables for layout, the expanded capabilities for typographic control afforded by CSS unquestionably boosted the ability of designers to create Web sites with layouts that focused on type. With CSS, designers can specify fonts and offer a list of alternatives, select a default font size and scale others relative to it, and use complex highlighting, underlining, and other effects. CSS offer control over many facets of typography simply impossible with only HTML. However, type was just the beginning. The decision to use a box model for the visual layout of page elements via CSS carried over design practices cultivated in the use of tables for layout. CSS treats design elements as rectilinear shapes with margins, padding, and borders—much like the `cellspacing`, `cellpadding`, and `border` attributes that can be added to `<table>` tags. For each element, width can be specified as an absolute number or a percentage relative to surrounding or parent elements. In other words, the CSS specification legitimized and extended practices that designers such as Powazek had previously used in unintended ways. Finally, CSS offered modularity: a small number of style sheets could be used to control an entire Web site and to change its design across the board with a single update (Zeldman 220–21). This form of templating moved Web design much closer to grid-based design because a single look and feel could be easily applied to many pages—as was the case with the pasteup boards and master pages of mechanical and computerized grid design.

The first CSS specification (CSS level 1) was finalized in late 1996, and CSS 2 followed in May 1998. But browser support for CSS was inconsistent for years afterward, slowing implementation of grids via CSS (Collison 235). Identical code would look very different in Microsoft Internet Explorer and its competitors, Mozilla, Opera, Netscape, and later Safari and Firefox. A potential resolution was found in CSS work-arounds such as Tantek Çelik's so-called box model hack, which used funky CSS code to trick Explorer 5 into visual rendering of CSS that matched other Web browsers.

But these and other hacks made designing with CSS difficult and severely undercut the "one code for many browsers" spirit of standards-compliant design often cited as the reason for ditching <table> in the first place. In fact, Peter-Paul Koch called them "a danger to Web development, both from a psychological and from a technical point of view." No surprise, then, that even as the standards-compliance movement gained momentum, use of tables for layout continued. Zeldman's *Designing with Web Standards* offered a hybrid model that combined CSS with tables for layout, and CSS guru Eric Meyer suggested that layouts that depended heavily on grid design should use <table>. In summary, tables did not disappear quickly, and grids marched on. Even after 2002, when the release of the Mozilla browser provided stronger support for sophisticated uses of CSS such as positioning, visual rendering of CSS was uneven, especially where the grid is concerned. That designers persisted in using the grid despite these issues shows the attractiveness of the international style.

Future development of CSS standards appears to favor layout grids as well. A model for grid positioning in CSS has been published as a W3C working draft, proposing that CSS add "capabilities for sizing and positioning in terms of a scalable grid" in order to facilitate adaptable multicolumn grids (Mogilevsky and Mielke). A similar proposal, the CSS Advanced Layout Module, works from the assumption that all pages are designed on flexible grids (Bos, "CSS Advanced"). It is unclear whether these proposals will be adopted or whether alternative methods that allow other methods of page layout will emerge. For example, Aaron Gustafson has proposed both rotation of objects and the addition of polygons to the rectilinear box model—common capabilities in page layout software. However, given that Bos has called for "a way to let designers specify their grid directly in CSS and position elements with reference" to that grid, "just like in tables" (Bos, "Device-Independent"), it appears that the design heritage of <table> will carry forward into its successor—perhaps not as the only way to create layouts on the Web, but at least one of several alternatives.

Grids Up Front, Grids Always

Although much of this essay has focused on the visual elements of grid design, the connection of the grid to greater control and better delivery of message is equally significant. As Ray Roberts notes:

> Grids are fundamentally about a way of thinking. They are used to help bring order to a page and to impose structured thinking into the design process.

> In making these decisions, a designer is generally helping to make content
> accessible. This engagement with the wider world is a political act. (19)

This sentiment was shared by many of the designers who made the international style part of the everyday visual lexicon. I have already noted the broad purposes promised for the grid by Müller-Brockman and his contemporaries, often summarized as favoring objectivity, rationalism, clarity, facilitation of commerce, and control. Designers openly proposed a connection between their clients' use of the international style and the achievement of these goals. For example, as the United States government adopted grids for many of its publications in the 1970s, perhaps most famously Massimo Vignelli's "Unigrid" program for the National Park Service, John Massey suggested his work would bring the Department of Labor

> uniformity of identification; a standard of quality; a more systematic and
> economic template for publication design; a closer relationship between
> graphic design (as a means) and program development (as an end) so that
> that proposed graphics system will become an effective tool in assisting the
> department to achieve program objectives. (qtd. on Meggs and Purvis 379)

For many designers, those were the goals of the grid. We can read the broad turn of Web design toward the grid as preference for the vision of the Web as presented by Jakob Nielsen over that of Ted Nelson. Its role as a controlling visual element restrained, even destroyed, the visionary, liberatory, experimental future of hypertext that Nelson imagined in favor of a Web centered on objectivity, order, and clarity. In this way, we see the opposition of grid to network Mark C. Taylor maps in *The Moment of Complexity*, where grids, like walls, "divide and seclude in an effort to impose order and control," as opposed to webs, which "link and relate, entangling everyone in multiple, mutating, and mutually defining connections in which nobody is really in control" (23). The preference for grids shows that designers sought to ally their clients with walls, boundaries, order, and control—all usually identified with the stability of the bricks-and-mortar manufacturing of the old economy, not the silicon and service of the new.

Notably, the same designers who saw the grid as an engine of standardization and rationality also saw the potential for trouble. Hurlburt was one of many who warned that grids can become too regular:

> When it is used with skill and sensitivity it can lead to the production of
> handsome and effective pages and it can give the overall design a sense of
> cohesion and continuity that has a distinctive unifying effect. However, in

the hands of a less able designer or when the priority is given to the structure, rather than the creative concept, the grid can become a straitjacket that
produces dull layouts and a rigid format. (18)

Timothy Samara memorably casts this problematic as "making and breaking the grid"—the title of his book, which argues that grids are more effective if design elements occasionally break the grid, or apply it flexibly by
alternating the number of cells used for given elements (headers, main
content, sidebars, etc.). Samara supports "breaking" or "deconstructing"
with a handbook of techniques, such as "spontaneous optical composition" and "conceptual or pictorial allusion." But little discussion about
creating flexible grids has taken place in Web design. In large part, the use
of <table> to implement grids makes impossible any of the "breaking"
methods Samara demonstrates—and which were regularly used by exem-
plars of the international style. Although page layout software allows, even
facilitates, deconstructed or broken grids by allowing designers to use irreg-
ularly shaped elements, variable transparency, and similar techniques, this
is not the case for tables, CSS, and Web authoring software.

Molly E. Holzschlag takes up this argument in "Thinking Outside the
Grid." With a nod to Samara, Holzschlag suggests that it is "the constraints
of the table-based layout that have kept us in visual gridlock for so long,"
codifying a rigid interpretation of grids that suffers from the negative effects
noted by Hurlburt. The result? A flat, regular grid, highly standardized across
a given site, but also little differentiated from other sites that use similar
standard widths, type sizes, and design best practices. Other designers have
criticized this homogeneity and uniformity as least-common-denominator
minimalism. Karl Stolley suggested the name "corporate-traditional," rif-
fing on Albert Kitzhaber's label for the staid, predictable rhetoric of col-
lege composition. Clark's proposed appellation, "international compliant
style," rightfully notes the importance of *compliance,* both in terms of fol-
lowing W3C standards, but also positioning it as the core part of the aes-
thetic, perhaps more important than typography or other elements of the
style. I want to suggest we read "international compliant" as the expres-
sion of a problem: the widespread (even global) collapse of design diver-
sity caused by preference for compliance over style.

Designers were comfortable not breaking the grid because they believed
that by doing so, they would more strongly project an ethos of control. But
as Alexander R. Galloway notes, that is not necessarily the case. In much
the same way that the international style gained its ethos of control from

the violations of its typographic grid, protocological systems of control gain force from dialectical tension:

> What contributes to this misconception (that the Internet is chaotic rather than highly controlled), I suggest, is that protocol is based on a *contradiction* between two opposing machines: One machine radically distributes control into autonomous locales, the other machine focuses control into rigidly defined hierarchies. The tension between these two machines—a dialectical tension—creates a hospitable climate for protocological control. (8)

And therein lies a great irony: by adapting the grid universally and never breaking it, Web designers failed to match their designs to the protocological systems of control that the history of tables for layout repeatedly demonstrates. Doesn't a design that retains its clarity and efficiency despite occasional (and carefully planned!) violations of its internal structure make a stronger argument for order and rationality than one that does not? That is not to say designers missed an opportunity to express strong support for protocol itself. By following the rise and decline of tables for layout from start to finish, we can see repeated and consistent approval of the work of protocol: in the adaptation of the grids of the international style to the Web; in the transfer of the aesthetics of layout via <table>, a non-standard practice, to the emerging standards of CSS; in the adaptation of modular CSS as an engine for decentralizing control of visual identity; and of course in broad support for standards-based Web design and its explicit ties to protocological systems of power distribution. More than anything else, this support for protocol was the "political act" Roberts names.

Still, I believe the political statement would have been far stronger had it been more nuanced. Like Galloway, I invoke protocol not to warn against it but to suggest that those of us who see Web design and/or standards compliance as political acts—and are committed to them—should continue political work via protocol. Of the Internet as a whole, Galloway writes, "Despite being a decentralized network composed of many different data fragments, the Internet is able to use the application layer to create a compelling, intuitive experience for the user" (64). I would make the same claim for standards-compliant Web design and for the adaptation of the international style whose history I have traced here. Like Holzschlag, I think Web design needs to open up the grids of <table> and loosen its rigid application of the international (compliant) style. The way toward that future includes modifying Web standards to better support a more flexible approach to the grid, developing best practices that offer strong alternatives to <table> and its descendants, and creating examples of designs that embody the values of the international style but not necessarily its visual approach. In summary, we need a Grid 2.0 that, like Web 2.0, is simultaneously a rethinking of the grid and a return to values characterizing its technical heritage.

Works Cited

Atwood, Jeff. "Let's Build a Grid." *Coding Horror: Programming and Human Factors.* 29 May 2007. Accessed 12 February 2009. http://www.codinghorror.com/blog/.

Bjørkøy, Olav, et al. "Blueprint: A CSS Framework." Accessed 12 February 2009. http:// code.google.com/.

Bos, Bert. "CSS Advanced Layout Module: W3C Working Draft 9 August 2007." 9 August 2007. Accessed 12 February 2009. http://www.w3.org/.

———. "The Device-Independent Browser: CSS and Grid Layout." 13 May 2005. Accessed 12 February 2009. http://www.w3.org/.

Cederholm, Dan. *Bulletproof Web Design.* Berkeley, Calif.: New Riders, 2006.

Çelik, Tantek. "Box Model Hack." 12 February 2009. http://tantek.com/.

Clark, Joe. "The IC-Style." *Fawny.blog.* 20 September 2003. Accessed 25 April 2009. http://fawny.org/blog/.

Cloninger, Curt. *Fresh Styles for Web Designers: Eye Candy from the Underground.* Berkeley, Calif.: New Riders, 2001.

Collison, Simon. *Beginning CSS Web Development: From Novice to Professional.* Berkeley, Calif.: Apress, 2006.

Galloway. Alexander R. *Protocol: How Control Exists after Decentralization.* Cambridge, Mass.: MIT Press, 2004.

Gustafson, Aaron. "Wouldn't It Be Nice?" *Easy Reader.* 27 June 2007. Accessed 12 February 2009. http://www.easy-reader.net/.

Holzschlag, Molly E. "Thinking Outside the Grid." *A List Apart* 209. 19 December 2005. Accessed 25 April 2009. http://alistapart.com/.

Hurlburt, Allen. *The Grid: A Modular System for the Design and Production of Newspapers, Magazines, and Books.* New York: Van Nostrand Reinhold, 1978.

Koch, Peter-Paul. "Keep CSS Simple." *Digital Web Magazine.* 6 November 2003. Accessed 20 March 2008. http://www.digital-web.com/.

Lawson, Bruce. "CSS Zen Garden Submission: 'Geocities 1996.'" 16 December 2004. Accessed 22 March 2008. http://www.brucelawson.co.uk/.

Lynch, Patrick, and Sarah Horton. "Alignment." *Web Style Guide.* 2nd ed. 2002. Accessed 12 December 2007. http://webstyleguide.com/.

MacManus, Richard. "The Evolution of Corporate Web Sites." *Digital Web Magazine.* 28 April 2004. Accessed 25 April 2009. http://www.digital-web.com/.

Meggs, Philip B., and Alston W. Purvis. *Meggs' History of Graphic Design.* 4th ed. Hoboken, N.J.: John Wiley & Sons, 2006.

Meyer, Eric. "CSS Gridlock." *Eric's Archived Thoughts.* 5 September 2004. Accessed 12 February 2009. http://meyerweb.com/eric/.

Middendorp, Jan. *Dutch Type.* Rotterdam: 010 Publishers, 2004.

Miner, Wilson. "Setting Type on the Web to a Baseline Grid." *A List Apart* 235. 9 April 2007. Accessed 17 December 2007. http://www.alistapart.com/.

MIT (Massachusetts Institute of Technology). Home page. *Archive.org.* 9 February 1997. Accessed 20 March 2008. http://web.archive.org/.

Mogilevsky, Alex, and Markus Mielke, eds. "CSS Grid Positioning Module Level 3: W3C Working Draft 5 September 2007." 5 September 2007. Accessed 22 March 2009. http://www.w3.org/.

Morkes, John, and Jakob Nielsen. "Concise, SCANNABLE, and Objective: How to Write for the Web." 1997. Accessed 12 February 2009. http://www.useit.com/.

Müller-Brockman, Josef. *Grid Systems in Graphic Design.* Sulgen: Niggli, 1996.

NBC (National Broadcasting Company). Home page. *Archive.org.* 22 October 1996. Accessed 20 March 2008. http://web.archive.org/.

NCSA (The National Center for Supercomputing Applications). Home page. *Archive.org.* 10 December 1997. Accessed 20 March 2008. http://web.archive.org/.

Niederst, Jennifer. *Web Design in a Nutshell: A Desktop Quick Reference.* Sebastopol, Calif.: O'Reilly, 1999.

Nielsen, Jakob. *Designing Web Usability: The Practice of Simplicity.* Berkeley, Calif.: New Riders, 2000.

———. "Inverted Pyramids in Cyberspace." *Jakob Nielsen's Alertbox.* June 1996. Accessed 25 April 2009. http://www.useit.com/.

Powazek, Derek M. "The Basic, Basic Table." *Webmonkey: The Web Developer's Resource.* 21 November 1996. Accessed 17 December 2007. http://www.webmonkey.com/.

Raggett, Dave, Jenny Lam, Ian Alexander, and Michael Kmiec. *Raggett on HTML 4.* Reading, Mass. Addison Wesley Longman, 1998.

Roberts, Ray. "The Principle of the Grid." In *The Designer and the Grid,* edited by Julia Thrift and Lucienne Roberts, 17–30. Hove: RotoVision, 2005.

Samara, Timothy. *Making and Breaking the Grid: A Graphic Design Layout Workshop.* Beverly, Mass.: Rockport Publishers, 2002.

Siegel, David. *Creating Killer Web Sites: The Art of Third-Generation Site Design.* Indianapolis: Hayden Books, 1997.

Smith, Nathan. "960 Grid System." Accessed 22 March 2009. http://960.gs/.

Spool, Jared M. "As the Page Scrolls." 1 July 1998. Accessed 22 March 2009. http://www.uie.com/.

Stolley, Karl. Electronic mail correspondence. 23 January 2006.

Taylor, Mark C. *The Moment of Complexity: Emerging Network Culture.* Chicago: University of Chicago Press, 2001.

Veen, Jeffrey. "Aesthetics for the Web: Lesson 2." *Webmonkey: The Web Developer's Resource.* 16 February 1998. Accessed 17 December 2007. http://www.webmonkey.com/.

Vinh, Khoi, and Mark Boulton. "Grids are Good (Right?)" *Subtraction.* 10 March 2007. Accessed 25 April 2009. http://www.subtraction.com/.

Weakley, Russ. "Liquid Layouts the Easy Way." *Maxdesign*. 30 December 2003. Accessed 20 March 2008. http://www.maxdesign.com.au/.

Williams, Robin, and John Tollett. *The Non-Designer's Web Book*. 3rd ed. Berkeley, Calif.: Peachpit Press, 2006.

Yahoo. Home page. *Archive.org*. 1 February 1997. Accessed 20 March 2008. http://web .archive.org/.

Zeldman, Jeffrey. *Designing with Web Standards*. 2nd ed. Berkeley, Calif.: New Riders, 2006.

Afterword

\<meta> CASUISTIC CODE

CYNTHIA HAYNES

\<meta> *tags can be found in the head of a document. An early application of* \<meta> *was the* http-equiv="refresh" *attribute, a way to instruct the browser to reset the Web page after a designated period of time. In addition to refreshing the page, Web writers could designate another URL as the destination point after the refresh, and thus cause the Web page to automatically change URLs.* \<meta>*'s best-known use is providing keywords, descriptions, and other metadata to search engines that used the content of the tags to build their indexes. However, this use is all but forgotten as search engines have turned to other means to build their indexes. Google, in particular, does not rely on* \<meta> *tags for fear of being manipulated, as director of research Monika Henzinger has stated. The move away from meta-based search is a response to spammers and "search engine optimization" consultants abusing the* \<meta> *tag, a practice known as keyword stuffing. Keyword stuffing inserted irrelevant information into* \<meta> *tags in order to bias search engine results. Recently, a common distinction between data and metadata—the former primary and visible, the latter ancillary and invisible—has blurred, as folksonomies, tagging, and other practices that revolve around user-created metadata have become popular.* \<meta> *has declined, then, not because metadata is less important than it used to be, but because its flexibility allows for new types of usages.*

> Code is the only language that is executable, meaning that it is the first discourse that is materially affective.
>
> **ALEXANDER GALLOWAY,** *PROTOCOL*

AT FIRST GLANCE, IT WAS A KILLING MACHINE. Not that I've seen that many. Like so many of the exhibits at the U.S. Holocaust Memorial Museum, knowledge and the *everyday* artifact re-presents an abject form of information lurking in the pins and punch cards that recorded details about *everyday* people. This particular machine was manufactured by the Deutsche Hollerith Maschinen Gesellschaft (Dehomag), a subsidiary of IBM. According to Edwin Black, thousands of these machines were situated in various camps during World War II in the camp's Labor Service Office, what prisoners at Bergen-Belsen called "the lion's den" (*IBM* 20). Hole punch categories on Hollerith cards detailed "nationality, date of birth, marital status, number of children, reason for incarceration, physical characteristics, and work skills. Sixteen coded categories of prisoners were listed in columns 3 and 4, depending upon the hole position: hole 3 signified homosexual, hole 9 for antisocial, hole 12 for Gypsy. Hole 8 designated a Jew" (21). Eventually, Black maintains, "the infamous Auschwitz tattoo began as an IBM number" ("At Death's Door").

This is code for casuistry of the worst kind. We know from Kenneth Burke that abstract principles applied to specific cases can sometimes lead to justifications for refining the principles themselves. Thus, Burke warned that "the process of casuistic stretching must itself be subjected continually to conscious attention. Its own resources . . . must be transcended by the explicit conversion of a method into a methodology" (232). The difference is that "casuistry as a method" is "the concealing of a

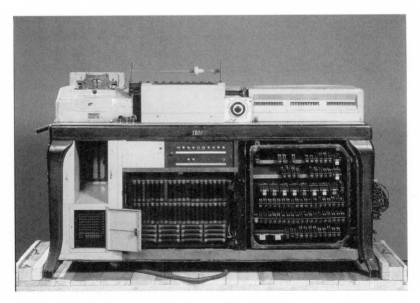

Figure A.1. Hollerith tabulation machine from IBM (USHMM #N00013). Courtesy of the United States Holocaust Memorial Museum.

strategy," while "casuistry as a methodology" is "a description of a strategy" (232). That said, before that day, I had generated a remixed tag cloud of this collection's chapter headings as a means of casuistic invention for an afterword. Each author, it seemed, had stretched the casuistic tale of their markup tag, beginning with Rice and Dilger's reminder that to explore keywords is somewhat of a dance with the devil, so to speak, during which the faint strains of the more nefarious practices of markup may be heard. In that spirit, and being attuned to their call, you can imagine the tag cloud building in my head as I stood in front of this machine. What I am about to stretch is the casuistic code of hole 8. What I hope to avoid is the refinement of a principle, which is itself a refined way of saying that at all costs one should avoid the twisted logic that uninstalls the ethics of technology and information control.

How to unplay the race card. Running the sequence now. If tagging is encoded at the very intersection of "international business" *solutions* and political expediency, then keypunch operators implemented (by proxy) the original `<meta>` tag, and sorting machines became the search engines ranking and optimizing the fate of millions of Jews. Hitler's algorithm installed a mass "state of exception" in the history of *categorical* violence (hole 8 designates Jew). According to Giorgio Agamben, "If the law employs the exception—that is the suspension of law itself—as its original means of referring to and encompassing life, then a theory of the state of exception is the preliminary condition for any definition of the relation that binds and, at the same time, abandons the living being to law" (1). In Hypertext Markup Language, the *suspension* `<meta>` tag marches lockstep into the ambiguous zone of the history of computing, which is at once both the (innocuous) history of human accounting and the (noxious) counting of humans as nonexistent. Alexander R. Galloway and Eugene Thacker raise the obvious implication of such ambiguity: "The question of nonexistence is this: how does one develop techniques and technologies to make oneself unaccounted for?" (135).

One way to understand the question is to situate it in the context of the rising need to control information in the late nineteenth century. Jan Rune Holmevik chronicled the history of modern computing in his 1994 thesis, *Educating the Machine*. Holmevik points to the emergence in the United States of the need to "find more efficient ways of handling and processing the vast amounts of census data" in years leading up to the 1890 census (Holmevik 38). Herman Hollerith, a Census Bureau engineer, devised "a new punched-card based electromechanical information processing and tabulating system" (38). This megamachine was soon adopted worldwide after the success of the 1890 census. By 1896, Hollerith "had established the Tabulating Machine Company," which changed its name in 1924 to International Business Machines (IBM) (39). What we now know about the commercial ties between IBM and the Nazis raises the question of how Hitler could so easily indenture a machine in the service of genocide. In Holmevik's study, he suggests that Lewis Mumford provides a clue:

> According to Mumford, as a social organizational principle, the megamachine originated in ancient Egypt, where the center of authority rested with one absolute ruler, the earthly representative of the Sun God. He believed that several megamachines had existed throughout history, but what distinguished the new megamachine of the twentieth century was that the center of authority had become the system itself. "The computer," he argues, "turns out to be the Eye of the reinstated Sun God, that is, the *Eye of the Megamachine*." (Holmevik 35; Mumford 274)

The relation between the divine and the machine became both codified and reified during the years of Hitler's emerging god complex. However, as Merry Madway Eisenstadt anticipates, executives at IBM are unwilling to draw a direct link between technology and genocide (much less between gods and machines). Eisenstadt acknowledges that "machines have no national allegiances and no moral code." Citing Michael Berenbaum's *The World Must Know,* she helps make the link explicitly visible: "The IBM technology was neutral; its use by the Nazi regime was malevolent. Clearly, its potential was understood by the German manufacturer," Berenbaum states, citing an exuberant statement by Dehomag director Willy Heidinger in 1934 about the future role of statistics and tabulation machinery in the Nazi state:

> We are recording the individual characteristics of every single member of the nation onto a little card. We are proud to be able to contribute to such a task, a task that makes available to the physician of our German body-social [i.e., Adolf Hitler] the material for his examination, so that our physician can determine whether, from the standpoint of the health of the nation, the results calculated in this manner stand in a harmonious, healthy relation to one another, or whether unhealthy conditions must be cured by corrective interventions. We have firm trust in our physician and will follow his orders blindly, because we know that he will lead our nation toward a great future. Hail to our German people and their leader!' (Eisenstadt)

Such `<meta>` tags for insanity chillingly mark the means by which Nazi megamachinery, forged in the factory of fascism, represents the pathological idolatry of indices—a collective index finger pointing inexorably to hole 8—the mark (and birth) of a malevolent search engine. If today's Web search engines are virtual descendants of the Hollerith machine, then the `<meta>` tag is the indexical means to achieve the optimal number of *hits*, or SEO, search engine optimization. It attempts to transcend the neutrality of data and installs metadata as the ultimate "quality control." Though not explicitly tackling metadata, Galloway and Thacker name the intersection of sovereignty and networks as *the* juncture of repression and control (5). *"This is why,"* they argue, *"contemporary political dynamics are decidedly different from those in previous decades: there exists today a fearful new symmetry of networks fighting networks. One must understand how networks act politically, both as rogue swarms and as mainframe grids"* (15). *"Connectivity is a threat. The network is a weapons system"* (16). In their analysis, Galloway and Thacker contend that resistance and intervention cannot succeed against "states of exception" that perfect (by adopting) terrorist network tactics. Such "asymmetrical" forms of conflict require an "exploit" (21). Because networks are "largely immaterial," they "operate through the brutal limitations of abstract logic (if/then, true or false). *Protocological struggles do not center around changing existent technologies but instead involve discovering holes in existent technologies and projecting potential change through these holes. Hackers call these holes 'exploits'"* (81).

The undeniable upshot of their solution is the shadow figure of hole 8. It was also an exploit. There is a thin line between *exploit* and *exploitation*. While I am not forcing a connection that Galloway and Thacker have ignored, I am raising the bar for their discussion of exploitation and asking that we include genocide as a "metasploit"[1] that is much more serious than their focus on the "practice of bioprospecting" and the exploitation of human labor (135). In other words, as a `<meta>` tag designed to optimize their theory of networks, Galloway and Thacker's *exploit* is inadequate to help the "impoverished classes" by exposing the "exploitative classes" they term "source fetishists," and McKenzie Wark calls "the 'vectoralists'" (135). The problem is that the exploit is a vector itself—metasploiting is about system "penetration testing" (Metasploit Project)—and you cannot penetrate a system without your own version of malicious code. As Nazi Germany became obsessed with statistics science, code provided the inner circle with names. Black reports that in his eagerness to please Hitler, Friedrich Zahn, head of the Bavarian Statistical office, gleefully summed up the government's new information power: "In using statistics, the government now has the road map to switch from knowledge to deeds" (*IBM* 59).

Shifting gears slightly, it is not unimaginable to consider Galloway and Thacker's exploit as a postprotocological *deus ex machina*. In Greek tragedy, this device was used as a means to supply a happy ending. For example, toward the end of *Medea*, Euripides staged a *deus ex machina* to solve the protagonist's crisis—literally speaking, a god was lowered onto the stage via some kind of machinery. Critics from Aristotle to Nietzsche have, however, criticized the literary device. In *The Birth of Tragedy*, Nietzsche explains the problem: "The *deus ex machina* has taken the place of metaphysical solace" (84). In essence,

> theoretical man. . . . fights against Dionysiac wisdom and art; it strives to dissolve myth; it puts in the place of metaphysical solace a form of earthly harmony, indeed its very own *deus ex machina*, namely the god of machines and smelting furnaces, i.e. the energies of the spirits of nature, understood and applied in the service of higher egotism; it believes in correcting the world through knowledge, in life led by science; and it is truly capable of confining the individual within the smallest circle of solvable tasks. (85)

The smallest circle is hole 8, and the god of smelting furnaces turned out to be the *deus ex machina* for the Nazis. The stage of their drama was a secret conference at an "elegant villa" located at Wannsee outside of Berlin (Black, *IBM* 366). According to Black, among the "senior Nazi leadership" who attended were several census and statistics experts who provided Adolf Eichmann with "a long list of Jewish populations" compiled by IBM's German subsidiary (366). "The conclave at Wannsee resulted in a Protocol, which outlined the massive demographic and geographic logistical challenge" (366).

This is not merely a cautionary tale. It is a narrative inflected by protocols, exploits, and database death squads. To casuistically stretch such code is to identify a neglected rhetorical history, but more importantly, it is to guard the question of *how* our field consorts with hacker culture and new media discourse. Lev Manovich explains that new media theories and practices are marked by the conflict between two dominant forms: narrative and the database. Manovich notes the historical predominance of narrative as the "key form of cultural expression of the modern age" (39) and argues that in today's computerized society, the database is the newest symbolic form with which "to structure our experience of ourselves, and the world" (40). So while computer games, for example, are certainly "experienced by their players as narratives" (41), the hidden data structure and algorithms necessarily project "the ontology of a computer onto culture itself" (42). In short, he writes, "any object in the world—[such as] the population of a city . . . is modeled as a data structure, that is, data organized in a particular way for efficient search and retrieval" (42–43). Manovich translates this new "cultural algorithm" into the necessity "to develop the poetics, aesthetics, and ethics of this database" (40).

The question has been put, as well and so well, by Steven B. Katz, who writes: "The question for us is: do we, as teachers and writers and scholars, contribute to this [ethics of expediency and technological] *ethos* by our writing theory, pedagogy, and practice when we consider techniques of document design, audience adaptation, argumentation, and style without also considering ethics?" (271). This is the casuistic <meta> tag without which search engine optimization remains merely an expedient perpetuation of metaphysical solace, our data *deus ex machina* logging in to the next server of genocide.

Note

1. The Metasploit Project, created in 2003, "provides useful information to people who perform penetration testing, IDS signature development, and exploit research. This project was created to provide information on exploit techniques and to create a useful resource for exploit developers and security professionals."

Works Cited

Agamben, Giorgio. *State of Exception*. Translated by Kevin Attell. Chicago: University of Chicago Press, 2005.

Black, Edwin. "At Death's Door: Archivist Finds IBM Site Near Auschwitz." 23 October 2008. Accessed 20 February 2009. http://www.scrapbookpages.com/ .

———. *IBM and the Holocaust: The Strategic Alliance between Nazi Germany and America's Most Powerful Corporation*. New York: Crown Publishers. 2001.

Burke, Kenneth. *Attitudes toward History*. 3rd ed. Berkeley: University of California Press, 1984.

Eisenstadt, Merry Madway. "Counted for Persecution; IBM's Role in the Holocaust." *Washington Jewish Week*. 17 September 1998. Accessed 26 February 2009. http://www.stockmaven.com/ibmstory.htm.

Galloway, Alexander R., and Eugene Thacker. *The Exploit: A Theory of Networks*. Minneapolis: University of Minnesota Press, 2007.

Holmevik, Jan Rune. *Educating the Machine: A Study in the History of Computing and the*

Construction of the SIMULA Programming Language. Trondheim, Norway: University of Trondheim Center for Technology and Society, 1994.

Katz, Steven B. "The Ethics of Expediency: Classical Rhetoric, Technology, and the Holocaust." *College English* 54, no. 3 (March 1992): 255–75.

Manovich, Lev. "Database as Symbolic Form." In *Database Aesthetics: Art in the Age of Information Overflow,* edited by Victoria Vesna, 39–60. Minneapolis: University of Minnesota Press, 2007.

Metasploit Project. 2003. Accessed 4 March 2009. http://www.metasploit.com/.

Mumford, Lewis. *The Myth of the Machine, Vol. 2: The Pentagon of Power.* New York: Harcourt Brace Jovanovich, 1970.

Nietzsche, Friedrich. *The Birth of Tragedy: And Other Writings.* Edited by Raymond Guess and Ronald Speirs. Cambridge: Cambridge University Press, 1999.

United States Memorial Holocaust Museum. 2009. Accessed 12 February 2009. http://www.ushmm.org/.

CONTRIBUTORS

SARAH J. ARROYO is associate professor of English at California State University, Long Beach.

JENNIFER L. BAY is assistant professor of English at Purdue University, where she teaches courses in professional writing, new media, and rhetorical theory. Her work has appeared in journals such as *jac* and *College English*, as well as in edited collections.

HELEN J. BURGESS is assistant professor of English at the University of Maryland, Baltimore County.

BRADLEY DILGER is associate professor of English at Western Illinois University.

MICHELLE GLAROS is associate professor of communication at Centenary College of Louisiana.

MATTHEW K. GOLD is assistant professor of English at New York City College of Technology and a faculty member in the Interactive Technology and Pedagogy Certificate Program at the CUNY Graduate Center.

CYNTHIA HAYNES is associate professor of English and director of first-year composition at Clemson University. She is coeditor of *High Wired: On the Design, Use, and Theory of Educational MOOs* (2001) and *MOOniversity: A Student's Guide to Online Learning Environments* (2000).

RUDY MCDANIEL is assistant professor of digital media at the University of

Central Florida. He is coauthor of *The Rhetorical Nature of XML: Shaping Knowledge in Networked Environments* (2009).

Colleen A. Reilly is associate professor of English at the University of North Carolina, Wilmington. Her research and teaching interests include writing and technology and open-access scholarly electronic publication and citation.

Jeff Rice is associate professor of English and director of the Campus Writing Program at the University of Missouri. He is the author of *The Rhetoric of Cool: Composition Studies and New Media* (2007), *Writing about Cool: Hypertext and Cultural Studies in the Computer Classroom* (2004), and a coeditor of *New Media/New Methods: The Turn from Literacy to Electracy* (2008).

Thomas Rickert is associate professor of English at Purdue University. He is the author of *Acts of Enjoyment: Rhetoric, Žižek, and the Return of the Subject* (2007).

Brendan Riley is associate professor of English at Columbia College Chicago.

Sae Lynne Schatz is a research associate with the Applied Cognition and Training in Immersive Virtual Elements Laboratory at the University of Central Florida's Institute for Simulation and Training.

Bob Whipple is professor of English and chair of the English department at Creighton University.

Brian Willems teaches literature and media culture at the University of Split, Croatia. He is the author of *Hopkins and Heidegger* (2009) and *Facticity, Poverty, and Clones: On Kazuo Ishiguro's "Never Let Me Go"* (2010).

INDEX

Abbott, H. Porter, 194
Achewood, 33
Acid Phreak (Ellias Ladopoulos), 18
Adorno, Theodor, 87–88
advertising, 98, 100, 107, 126; writing
	style of, 216, 218
aesthetics, xx, xxii, 54, 87, 102, 139, 197,
	234; of linearity, 126–27; multimedia
	and, 112, 114–15, 117, 120; ornamen-
	tation and, 142–44; style and, xix,
	73–74, 77, 215; visual, 207–8
Agamben, Giorgio, 230
aggregation, xii, 63–64, 145, 147n3
Ajax, xii, xxii, 65n3, 183, 192
Alexander, Jonathan, 146
algorithms, 81, 89, 174, 179, 183; culture
	and, 230, 234; search engine, 1, 49
Americans with Disabilities Act (ADA),
	34, 37, 42, 45n1
America's Next Top Model, 160–61
Andreessen, Marc, 186
animation, 98–99, 103, 108, 128, 140–42,
	187, 191
anonymity, 162–64
apparatus, 56, 170, 215
apparatus shift, xiv
architecture: disability and, 41; of Las
	Vegas Strip, 106–7, 108; ornament
	in, 141
Arvidsson, Adam, 156, 160–61

associative logic, xviii, 1–2, 27–28, 146–47,
	215
attention, 74
avatars, xxi, 161, 188–89, 190, 196

Barabási, Albert-László, 50, 53
Barthes, Roland, xxii, 128, 189, 193–96
Berners-Lee, Tim, xii–xiii, 4, 8–11, 14, 54–
	55, 70, 128, 215, 217; CERN (Organisa-
	tion Européenne pour la Recherche
	Nucléaire) and, 8–9, 215
Bérubé, Michael, 42
Bishop, Chris, 89
Blade Runner, 5
blogs, 52, 154–55, 161–63, 179–80; anony-
	mously authored, 163; WordPress
	and, xiv, 21, 51, 167, 187. *See also*
	microblogging
Bloomer, Kent, 141, 144
bodies, xxi, 51–53, 64–65, 82, 85–86, 88,
	150–65; code and, 150–51; human,
	119, 121, 150; keywords and, xiv–xviii;
	materiality of, 152–55; mediation,
	153–55, 164; mind and, 4–5, 7, 24–26,
	164n1 (*see also* Descartes, René);
	moving, 107; pedagogy and, 56–58; of
	Web page, 4, 13, 104, 150–52
Boler, Megan, 151
Bolter, Jay David, 60–62, 120–21, 126, 138
Bos, Bert, 220, 221

Brereton, John, 56

Brown, Denise Scott, xx, 106–7

browsers. *See* Web browsers; *names of specific browsers*

Burke, Kenneth, xvi, 229

Burnett, Ron, 29

Bush, Vannevar, 146

Cailliau, Robert, 10, 19n4

Cascading Style Sheets (CSS), xix, xxi, xxii, 67, 71, 81, 83, 112, 192, 217; content and, 154, 178; development of, 68–71, 174, 213, 215, 219–21, 224; typography and, 215–16; work-arounds and, xii, 220–21. *See also* markup; style

Castro, Elizabeth, 83

casuistry, xvi, xxiii, 229, 230, 234

CERN (Organisation Européenne pour la Recherche Nucléaire). *See* Berners-Lee, Tim: CERN

cinematic language, 116–18, 146; of D. W. Griffith, 116; of Soviet Montage, 117

Clark, Joe, 215, 223

Cloninger, Curt, 218, 219

Clough, Patricia Ticineto, 8

code: xii–xiv, xvi, xxi, xxiii, 73, 76, 82–86, 150–54, 163–64, 221, 229–30; accessibility and, 37–41, 44–45; bar, 85; blended with content, 89–94; blended with images, 129; commenting, 184–85; culture and, 1–2; execution of, 175–78; identity and, 157–58; images and, 191; implicit meaning and, 29–30; materiality of, 151; moral, 231; protocol and, 16–18; punctuation and, 82–83; style and, xix, 68–70; visual rendering of, 180. *See also* hacking; HTML; markup

collaboration, 27–28, 31, 77, 180

collective wisdom, 28, 51

college composition, xix, 55–56, 223. *See also* English A

Connolly, Daniel, 128

content management systems (CMS), xiv, 51, 179, 181–83, 184, 219

Coover, Robert, 144

critique, xiii, 12–13, 55, 57, 60

Danger Mouse (Brian Joseph Burton), 19n7

databases, 11, 30, 51–52, 168, 176–83; logic of, xxii, 179–80, 182–83, 233–34 (*see also* Manovich, Lev)

dating Web sites, 154, 160–61

Davis, D. Diane, 25

Delicious (bookmarking system), 49, 51, 63–64

Derrida, Jacques, 5, 7, 101, 145

Descartes, René, 5, 7, 151, 164n1

design, xiii, xvii, xxii–xxiii, 29–30, 68–70, 78–79, 114, 131–32, 142–44, 182–84, 208–9, 215–19, 222–23; accessible, 40–45; conceptual, 23–24, 31; database, 184; of newspapers, 137–38, 139; variation in, 23, 126–27, 214; vulgar or flashy, 76, 102, 106–8, 143–44, 174. *See also* grids; Web design

deus ex machina, 233, 234

Dibbell, Julian, 155

différance, 94, 101

Dreamweaver (software), xiv, 76, 84, 168, 213

Drupal, 51

Duchamp, Marcel, 114

Dyens, Ollivier, 115, 116, 118

education, xix, 7, 18, 26, 49, 53, 55–62, 65, 158

Eisenstein, Elizabeth, 106, 114

Eisenstein, Sergei, 116–17

ekphrasis, 193

electric light, 100–104. *See also* McLuhan, Marshall

electronic writing, xx, 27, 38, 60–61, 112–22, 185. *See also* Web writing

electracy, xiv–xv

Engelbart, Douglas, 111

English A, 53, 55–60, 62–64
errors, 40, 83–84, 90–92, 105
exemplars (printing), 170, 173. *See also*
  pecia
exploits (software), 17–18, 19n7, 232–33

Facebook, xii, xxi, 51–52, 154, 157–59,
  161, 165n6
film, xvi, xx, 1–2, 5, 16, 93, 103, 111, 116–
  17, 120, 122, 146, 204, 215
Firefox, 101, 104
Flash (browser plug-in), 100, 108
Flickr, 51–52, 63, 65, 186, 208–9
folksonomy, xviii, 51, 144, 228
Fuller, Matthew, xv–xvii

Galloway, Alexander R., xvii, 10–12, 15,
  17, 58–59, 223–24, 228, 230, 232
gender, 151, 158, 161, 163–64
Gibson, William, 151; *Pattern Recognition,*
  xix–xx, 93–94
Goldsmith, Oliver, 133, 136
Gombrich, E. H., 142–43
Google, xii, 1, 3, 50, 158, 228; Maps, 187
Graff, Gerald, 58
grammatology, xiv, xvii
grids (design), xxii–xxiii, 127, 129–31;
  dividing text and, 129; rationality and,
  127. *See also* design; Web design
Griffith, D. W., 116–17
Grosz, Elizabeth, 152

hacking, 17–18; culture of, 73, 233–34.
  See also code; markup; programming
Haig, Andrew, 61, 62
Hamacher, Werner, 82–83
Hamann, Johann Georg, 14
Hansen, Mark B. N., 152
Hardt, Michael, 8, 15
Harvey, David, 147n3
Hawhee, Debra, xviii, 24–26
Hayles, N. Katherine, xix, xxi; electronic
  literature and, 115, 118, 121; flickering
  signifier and, 91–92; invisibility of code

and, 168–69; materiality of coding
  and, 152–53
Hebdige, Dick, 77
Helfand, Jessica, 143–44
hermeneutics, 2
hidden intellectualism, 58
Hobbs, Catherine, 25
Hollerith machine, 229, 231, 232
Holmevik, Jan Rune, 230, 231
Holzschlag, Molly E., 223
HTML (Hypertext Markup Language),
  xix, 4, 22–23, 38, 67, 68, 98–99, 140,
  150, 168, 173–74, 175–79; specifications xiii, 4, 19n1, 38–39, 128, 174,
  186, 214; structural orientation of,
  68, 127–28; tables, 174. *See also* code;
  hypertext; markup; World Wide Web
  Consortium
HTTP (Hypertext Transfer Protocol), 12
Hurlburt, Allan, 222–23
hyperlinks, xviii, 28, 44, 57, 62–63, 99,
  117, 188, 190, 203, 208–9; continuity
  and, 116–17, 222; as generative, 54–
  55; networks and, 8–9, 49–52; as relationship, 55. *See also* hypermedia;
  hypertext
hypermedia, 61, 199, 209–10; as extension of text or hypertext, 128, 139–40;
  place and, 202–3. *See also* hyperlinks;
  hypertext
hypertext, xiv, 49–50, 55, 59–62, 117–18,
  122, 150, 175, 187, 196; divisions
  in, 126, 128; fluidity of, 190–92; as
  liberatory, 146, 222; linearity and,
  144–45; as multimedia, 71; as paradigm, 60; protocol and, 11; study
  of, 200; writing and, 60–61, 113–15.
  See also HTML; HTTP; hyperlinks;
  hypermedia

identity, xii, xix, 85–88, 92, 151, 155–58,
  164, 188, 195–96; authenticity of, 156–
  57, 159; blogs and, 161–62; fragmentation of, 155; human face and, 163;

individual, 58; play, 9, 155–56; theft of, 157, 165n5; visual, 219, 224

imageability, 203–5

images, xi–xii, xxi, xxii, 62, 67, 74, 93, 125, 131, 150–51, 174, 184, 186–99, 203–10, 216; accessibility and, 33, 34, 38–40; animated, 125, 140–44, 146; detail of, 198–99; focus on, 198; inferior to text, 29–30; mixed with text, 50–51, 61, 62, 74, 114–16, 120; online profiles and, 158–61; recognition of, 90–91; speed of, 198, style and, 77. *See also* Flickr; image schemata; imagetext; photography

image schemata, 208–9

imagetext, xxi, 128–29, 189–92. *See also* images; Mitchell, W. J. T.

indexing, xv, 1, 86, 228, 232

individuation, 7, 28–29, 53–54, 57, 61

installation art, 118–20, 152, 153

interactivity, 10, 75, 150, 215; of hypertext, 49–50, 61

international style, 214, 215–17, 219

Internet Explorer, 70, 98, 100, 104, 220

Izenour, Steven, xx, 106–7

James, William, 146–47

Jenkins, Henry, 26–27, 31n1

Johnson, Mark, 208

juxtaposition, 62–63, 112, 117; of typographic styles, 138

Kant, Immanuel, 5, 7, 13–15

Katz, Steven B., xxiii, 84, 86, 234

Kaycee Nicole (Web hoax), 156–57

Kendall, Lori, 156

Kennedy, Helen, 162, 164

keyword stuffing, 228

Kinnane, Ray, 61, 62

Kirschenbaum, Matthew G., xv

Kitzhaber, Albert, 57, 223

Klensch, Elsa, 77

Koch, Peter-Paul, 221

Kress, Gunther, 71, 78

Kurzweil, Ray, 90

Landow, George P., 50–51, 60–62, 113, 128

Lanham, Richard, 74

Latour, Bruno, xix, 52, 59, 62, 182

legibility (recognition of images), 205–6

Lessing, G. E., 120

lexia, 128

literature, electronic. *See* electronic writing

links. *See* hyperlinks

Lions' Commentary on Unix, 184

Lipson, Carol S., 131

literacy, xiv–xv, xx, 56, 70, 76; visual and, 23, 71, 78–79, 168, 184. *See also* print; Web writing; writing

Liu, Alan, xxiii, 53–55, 60

Lynch, Kevin, 203, 205

Lyotard, Jean-François, 27–28

Mandel, Barrett J., 6

Manovich, Lev, xxiii, 146, 147n3, 182–83, 186; database logic, xxii, 179–80, 182–83, 233–34; *The Language of New Media*, 120; logic of selection, 75–76

Manuel, Jay, 160–61

markup: age of, xi, xiv; best practices, 22, 169, 220–22; as dynamic, xiii–xv; form and content division in, 70, 174; of human bodies, 150–52; invisibility of, 167–69; nesting of, 178; rendering of, 174; structural vs. presentational, 174–75. *See also* code; hypertext; tagging, tags; *names of specific markup languages*

mashups, xii, 16, 19n7, 183

materiality, 24–25, 152–54

McGann, Jerome, 169

McKenzie, D. F., 169

McLuhan, Marshall, xi, xii, 54, 56–57, 60, 70–71, 101

memex, 146

metadata, xi–xii, xvii, xxiii, 1, 2–4, 228, 232. *See also* folksonomy; tagging, tags

method, xv–xvi, 59, 64, 89–90; casuistry and, 229–30; writing and, 71–72, 74–75 microblogging, 145, 147n4. *See also* blogs

Microsoft Internet Explorer. *See* Internet Explorer

Mitchell, W. J. T., xxi, 129, 189–90, 194–95

Montero, Maria, 198–99

Montulli, Louis, 102

Mosaic (browser), 106, 150, 186, 216

Moses, Myra, 84, 86

Müller-Brockman, Josef, 215, 218, 222

multimedia, 23–24, 26, 99, 103–4

MySQL, 167, 176–78, 180

Nancy, Jean-Luc, 86–87

Narmer Palette, 131

narrative, xx, xxii, xxiii, 4, 21, 115–17, 119, 131, 146, 180, 186–200, 203–5, 207–21; community, 187; databases and, 179–80, 233–34; dominance of, 115; images and, 193–99; markup and, 169; place and, 192, 205–7

National Federation of the Blind (NFB) v. Target Corporation, 43

Negri, Antonio, 8, 15

Nelson, Ted, xiv–xv, 49–50, 52, 59, 146, 186, 222

Net.art, 118, 120

Netscape, 1, 49, 98, 101, 102, 106, 111, 214, 220

networks, xvii, xviii–xix, xxi, 2, 7, 9–10, 27–30, 49–50, 61–65, 128, 144–45, 183, 188, 192, 199, 222, 224, 232; Facebook and, 165n6; neural, 89–91; political control and, 12–14, 18, 232–34; social, 53–54, 58–59; writing and, 115. *See also* social network services; Web

newsgroups, xviii, 34–37; Backbone Cabal and, 35

newspapers, xi; design of, 137–39, 219

Nielsen, Jakob, 117, 218–19, 222; alt attributes and, 34–35, 39–44; functionality and, 74, 98, 102, 107–8, 213–14; HTML frames and, xx, 112

Ning, 29

nodes (network), 9, 59, 155, 161, 203, 205–7, 210

Norman, Donald, 73–74

Nunes, Mark, xxii, 188, 192, 199–200

Ong, Walter, xiii, xiv, 114, 170

Opera (browser), 70

O'Reilly, Tim, 51

ornamentation, xx, xxi, 22–23, design and, 140–43, 144; print and, 131–32; style as, 73, 78

paradigms, 15, 75, 90, 153, 182, 201; hypertext and, 60; writing, 55

pattern recognition, xix, 82, 89–92, 95n3; neural networks and, 90–92; noise and, 91. *See also under* Gibson, William

pecia system (book production), xxi, 170–73; marks, 171–73, 179, 184–85

pedagogy, 24, 146. *See also* education

persuasion, 2, 15, 16, 106–7

photography, 158, 186, 189–90, 199, 216

PHP, 52, 175–79

phpMyAdmin, 180–81

place, xxii; collective or shared sense of, 196, 200, 203–4; narrative, 192, 205–7; as psychological construct, 200–202; space and, 118–22

plagiarism, xvii, 2, 16–18

poststructuralism, 7–9, 61

Powazek, Derek M., 214

print, xi, 27–28, 55–58, 92, 214–15; conventions of, 99, 106, 108, 114, 218–19; divisions in, 132, 136, 137; errors in, 105, 173; influence on electronic text, 108, 115, 118, 218–19; ornamental features of, 131–34; production of, 170–73. *See also* literacy; writing

profile (online), xxi, 52, 158, 160, 197

programming, xiv, 73, 91, 118, 182–83. *See also* code; hacking; markup

protocol, xvii, xix, 9–19, 58–60, 192, 215, 232–33; contradictions in, 2, 224;

control and, 12–15; education and, 58; layered nature of, 12–13, 14, 58–59; rhetoric and, 15–18; Web standards and, 11, 68–69

Pugin, A. W., 142–43

punctuation, 56–57; as code, 82, 94n1

Raymond, Eric, 73

readymades (art), 114

Reid, Brian, 35–36

Reilly, Elizabeth, 131–32

remediation, 121, 126, 138–39, 141

rhetoric, xxi, xxiii, 10, 13, 21–22, 29, 50, 64, 68, 146, 151, 154–56, 161–64, 187, 206, 233; of cinema, 115–16; delivery, 31n2; first-year composition and, 55–56, 223; human bodies and, 24–27; images and, 199, 203–4, narrative and, 193–94, 196–97; of Las Vegas, 106–8; of markup, xii–xvii; pedagogy and, 146; protocol and, 15–18; style, 70, 73, 77

rhizome, 9; as structure of Internet, 10–11

Rhodes, Jacqueline R., 146

Rush, Sharron, 42, 44

Safari (browser), 101, 104

Samara, Timothy, 217, 223

Sample, Mark, 147n4

Sayre, Kenneth, 95n3

search engine, xxiii, 1, 3–4, 49–50, 228, 230, 232; algorithms, 1, 49; optimization for, 4, 232, 234. *See also* Google

security, 167, 192

server-side scripting, 167, 168, 175–79, 183–85

Shirky, Clay, 62–63

Siegel, David, xx, 125, 127–28, 138–39, 145, 213, 216

signage, 100, 102–5, 106–7

Sim City, 208

The Sims, 208

Slashdot, xii, 49

Slatin, John, 42, 44

social media, 51–52, 60–64. *See also* social network services; Web 2.0; *names of specific media*

social network services, xii, xvii, 29–30, 144, 154–61, 163–64; authenticity and, 159–61; friendship on, 159–60; profiles on, 158; statuses on, 159. *See also* Facebook; Flickr; Twitter; Web 2.0

software, 64–65, 76, 81, 83–84, 91–92, 152, 167–68, 173, 180–81, 183–84, 216, 219, 221, 223; accessibility and, 40–41; social, 51–52, 60–64. *See also names of specific software*

software studies, xv–xvii

SPIME, xix, 85–88, 95n2

Spinuzzi, Clay, 42–43

standards. *See* HTML: specifications; Web standards

state of exception, 230, 232

Sterling, Bruce: biot, 88–89; SPIME, xix, 85–88, 95n2

Stokols, Daniel, 198–99

Stolley, Karl, 223

Structuralism, 2

Strunk, William, 67, 76

style, xix, 23, 25, 67–68, 131, 140, 143–44, 150, 154, 161, 179, 198, 207; efficiency and, 73–74, 214; fashion, 77–78; guides, 71–72; international, xxii–xxiii, 214–18; learning, 23, 25; minimalist, 75, 142; separated from content, 67–68; typographic, 137–38, 215–17, 219–20. *See also* Cascading Style Sheets

subjectivity, xvii, 5–7, 9, 86–88; and human bodies, 150–52, 163–64

Suck (Webzine), 98

Suderburg, Erika, 119

tagging, tags, xi, 63–64, 167–68, 228, 230; markup and, xi, 151; metadata and, xvii, xxiii, 122, 180, 186, 230–33. *See also* code; folksonomy; markup; *names of specific markup languages*

Taylor, Mark C., 222

Thacker, Eugene, 58–59, 230, 232
Ticketmaster, 49
Tilton, Eric, 23
Tobias, Dan, 22–23
Tofts, Darren, 61, 62
Tuan, Yi-Fu, 201
Turkle, Sherry, 155
Twitter, 51, 145
typography, xi–xii, xxii, 70, 131–32; style
    and, 137–38, 215–17, 219–20

Ulmer, Gregory L., xiv
usability, 40–44, 102, 197; analyses of, 41–
    44; design and, 73–74; writing and,
    218. *See also* Nielsen, Jakob
Usenet, xviii, 34–37; Backbone Cabal and,
    35

Venturi, Robert, xx, 106–7
Victorian Web, 113–14
Viegas, Fernanda B., 162
View Source, xiii, xxii, 167–69, 175, 178, 183
Vignelli, Massimo, 222
virtual, xviii, 8, 10, 180, 184, 192, 200, 210,
    232; community, 202–3; opposed to
    material, 22, 24–27, 31; place, 202–8,
    210; space, 126–27, 131, 144–46, 186,
    195, 197
visual culture, xviii, 100; American, 100–
    102, 106–8

W3C. *See* World Wide Web Consortium
Wajnryb, Ruth, 195–96
wayfinding, 206–7
Web 2.0, 31, 51, 122, 157, 163–64, 168, 183
Web (World Wide Web), xi, xii, 8–10, 14,
    16, 19, 38, 54, 65, 67, 73, 75, 78, 83,
    106, 138, 140, 144, 174, 187, 215; text-
    only, 41–42, 102, 216. *See also* Web
    authoring; Web browsers; Web design;
    Web standards; Web writing
Web authoring, 76–77, 83–84, 98–99, 213,
    223. *See also* Web writing
Web browsers, xii, 3, 23, 102, 104, 139,

173–76, 178; assistive technology and,
    33, 40; bookmarks, 63 (*see also* Deli-
    cious); development of, 10, 106, 186,
    188, 216; HTML frames and, 111–12;
    standards and, 67–70; variation
    between, 101–2, 111, 127–28, 150,
    191–92, 214, 220–21; View Source, xiii,
    xxii, 167–69, 175, 178, 183; wars, 98.
    See also names of specific browsers
Web Content Accessibility Guidelines
    (WCAG), 38–40
Web design: 41, 44, 72, 83, 167–68; early,
    98–99, 102, 125–26, 138–40, 213–14,
    216; minimalist, 216, 219; multimedia
    nature of, 23–24, 143–44; need for
    professionalism in, 68, 81, 83, 102,
    112, 126, 143–44, 197; parallels to
    print, 218–29; standardization of,
    218–19, 223–24. *See also* design; grids
Web hosting, 125; for photography, 186
    (*see also* Flickr)
Weblogs. *See* blogs
Web standards, xvii, 14, 68–70, 76, 174,
    215, 220–21, 224; Web Content Acces-
    sibility Guidelines (WCAG), 38–40. *See
    also* World Wide Web Consortium
Web writing: conventions of, 113–15, 117;
    databases and, 180–83; difference
    from print, 144, 146; fluidity of, 191;
    functionality and, 112, 218; GUI
    (graphical user interface) and, 168,
    178–79, 180; linearity of, 139, 144–45;
    narrative in, 118; spatiality in, 118,
    120–22, 126–27; visual nature of,
    23–24, 126, 143, 190–92. *See also*
    electronic writing; Web authoring
Wegenstein, Bernadette, 153–54, 157, 158
Weinberger, David, 21, 30
Whittier, John Greenleaf, 132, 134, 135
Wikipedia, 21, 95n2
Williams, Joseph M., 72
Williams, Raymond, xv
windows (interface element), 3, 111, 112,
    131